INTRODUCTION TO COMPUTERS, STRUCTURED PROGRAMMING, AND APPLICATIONS

Module

A

Applications and Algorithms in Science and Engineering

INTRODUCTION TO COMPUTERS, STRUCTURED PROGRAMMING, AND APPLICATIONS

Module

A

Applications and Algorithms in Science and Engineering

C. WILLIAM GEAR
University of Illinois
Urbana, Illinois

SCIENCE RESEARCH ASSOCIATES, INC.
Chicago, Palo Alto, Toronto, Henley-on-Thames, Sydney, Paris, Stuttgart
A Subsidiary of IBM

Compositor	Advanced Typesetting Services
Acquisition Editor	Robert L. Safran
Project Editor	Jay Schauer
Special Editorial Assistance	Stephen B. Chernicoff
Text Design	Judy Olson
Cover Design	Michael Rogondino

LIBRARY OF CONGRESS CATALOGING IN PUBLICATION DATA

Gear, Charles William.
Applications and algorithms in science and engineering.

(His Introduction to computers, structured programming, and applications)
Includes index.
1. Science—Data processing. 2. Engineering—Data processing. 3. Algorithms. I. Title.
II. Series: Gear, Charles William. Introduction to computers, structured programming, and applications.
Q183.9.G4 502'.8'5 77-25545
ISBN 0-574-21188-8

10 9 8 7 6 5 4 3 2 1

Contents

Preface

This version of Module A (Algorithms and Applications) is intended for use by science and engineering students. Using the programming principles and informal language introduced in Module P (Programming and Languages), a variety of techniques and methods of solution are developed that are useful in wide classes of numerical and nonnumerical problems of interest to scientists and engineers. The material is organized so that later chapters depend minimally on earlier ones, to allow the instructor flexibility in the selection and ordering of topics. To help the instructor, a diagram appears at the beginning of this module, showing specific prerequisites for each chapter.

For a one-semester course, Module A can be used to amplify the material in Modules P and C by selecting example applications of interest to the students. Alternatively, Modules A, C, and P together provide enough material for a two-semester sequence in computer programming and applications. The first semester can cover the first six chapters of Module A along with most of Modules P and C and a language module; the remainder of Module A can be taught in the second semester, along with a second language if desired. The language modules are tied so closely to Module P that few additional ideas would have to be introduced: essentially it would only be necessary to provide exercises in the syntax of the second language. In a two-semester sequence, Chapters P10 and P11 could also be left for the second semester.

Module A can also be used in a separate course in computer applications for science or engineering students with adequate background in a nontrivial programming language. The informal language used to describe algorithms is natural for anyone with moderate exposure to a structured language; for students with experience only in Fortran or Basic, a quick review of the General Introduction and a brief discussion of the basic structured language constructs would be needed.

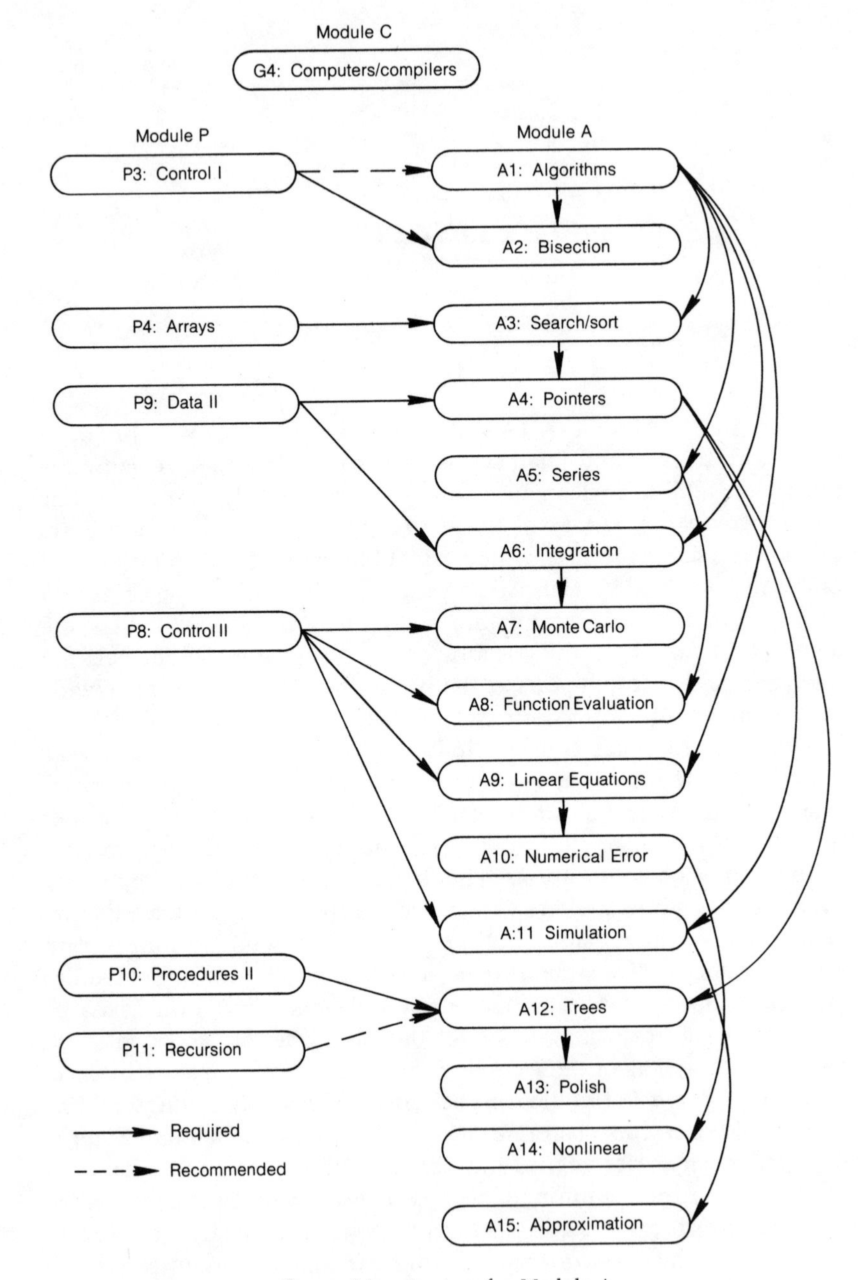

Prerequisite structure for Module A

Module

A

Algorithms and Applications in Science and Engineering

Before we can write a program, we must design an algorithm to solve the problem at hand. The design of the algorithm depends strongly on the structure of the data. For example, the only way to search for an item in an unordered list is by some form of sequential search, in which each item is examined in turn. However, if the list is ordered—alphabetically, for example—we can apply more efficient procedures. Accordingly, we might consider sorting an unordered list before searching it. We must decide whether the time needed to sort the list is a reasonable trade for the time saved in searching it. If only a few items are to be looked up, it will take more time to sort the list than will be saved on the searches; but if many items are to be looked up, the saving can be considerable. The design of good algorithms must take into account such considerations as whether a change in the data structure may lead to a better method of solution.

Computer applications arise in many diverse areas of activity. In business, computers are often used for *data processing,* manipulating large amounts of data representing company or government files, updating those files, and generating summary reports and records of individual transactions (weekly pay slips, orders, invoices, airline tickets, and so forth); *information retrieval,* which allows managers to examine the status of company or government files stored in the computer and spot potential trouble areas quickly; and *simulation,* the manipulation of numerical models of the real world. Scientific and engineering applications include the approximate solution of numerical problems, correlation and comparison of large amounts of experimental data, and information retrieval applied to textual data.

Many application areas are concerned with common problems and methods of solution. Early chapters in this module will examine some simple problems common to many areas and develop some important methods of solution. Later chapters will then address more specialized problems from particular application areas.

Chapter A1

Types of Algorithm

There is no universal set of rules for designing algorithms: each new problem may need a totally new approach. Indeed, it is this aspect of computer programming that can be the most pleasurable, providing a challenge akin to a crossword puzzle or chess problem and giving an outlet to the ingenuity and creativity of the programmer. There are, however, a number of basic *types* of algorithm that can frequently be used to solve a particular problem.

Five common types of algorithm are given below, followed by examples and a discussion of each:

- *Direct computation*—in which the exact answer is obtained by a sequence of elementary computations.
- *Enumeration*—in which all possible "answers" are tried in order to find one that solves the problem.
- *Divide and conquer*—in which the problem is divided into similar but smaller problems that can either be solved directly or be further subdivided by the same technique.
- *Iteration*—in which a series of increasingly precise approximate answers are computed until one is obtained that is "close enough." (An exact solution would require an infinite number of operations.)
- *Trial and error*—a type of iteration in which each successive approximation is based on the degree of error in the previous approximation.

Direct computation. The income-tax computation of Chapter G2 is an example of direct computation. This form of solution is applicable to simple problems in which the problem description itself specifies the computation needed to solve the problem.

Enumeration. A sequential search is an example of enumeration: each entry in a list is checked to see if it is the one sought. Enumeration is usually very slow, but sometimes it is the only method available. Often it is possible to start with an enumeration method and improve it by avoiding obviously impossible cases, as the following example shows.

Example A1.1 *Prime Numbers*

Given a positive number N greater than 1, find the smallest integer M > 1 that divides N exactly.

If the smallest divisor is N, then N must be prime. An enumerative method for solving this problem is simply to test each integer less than N, starting with 2, to see if it divides N. If one is found, it is the smallest. To program this solution, we need to be able to test whether M divides N. This is a basic operation in some computers and programming languages, but not in others. However, it can be programmed in terms of more elementary operations by testing whether N is equal to (N ÷ M) * M (see Chapter P2). Our first attempt at this program is shown in Program A1.1. If N is prime, the loop is executed N − 2 times (for the values M = 2, 3, . . . , N − 1). A little thought reveals that if N is not prime, one of its divisors must be less than or equal to the square root of N, so there is no need to test any values above that. Program A1.1a gives a revised version. For the case N = 127, Program A1.1 executes its loop 125 times, whereas Program A1.1a executes its loop only 10 times. A further improvement is possible by checking only for M = 2 and the odd numbers between 3 and the square root of N.

Program A1.1 *Find a divisor of* N

```
SMALLEST__DIVISOR: program
    integer M,N
        M←2
        do while M*(N÷M)≠N
            M←M+1
            enddo
        output M
    endprogram SMALLEST__DIVISOR
```

Program A1.1a ***Improved divisor program***

```
SMALLEST__DIVISOR: program
    integer M,N
        M←2
        do while M↑2≤N and M*(N÷M)≠N
            M←M+1
            enddo
        if M↑2>N then M←N endif
        output M
    endprogram SMALLEST__DIVISOR
```

Enumeration methods are the basis of many programs for nonnumerical problems, but because they can be so slow, it is essential to conduct a careful analysis to look for improvements.

Divide and conquer. Breaking a problem into simpler subproblems is a very powerful technique, useful for both numerical and nonnumerical problems. We will illustrate it with a search in an ordered list, such as a telephone book. One way of doing such a search is to open the book in the middle and see whether the item sought is before or after the middle entry. This can be done by a single comparison with the middle entry, because it is known that all items before that entry are alphabetically less and all items after it are alphabetically greater. Thus in one step we have reduced the size of the list to be searched by half. The same technique can now be applied to the smaller list. Thus, if the original list had 16 items in it, the first comparison leaves us with a list of 8 items to consider, the second with a list of 4, the third with a list of 2, and the last with a list of 1. A list of one item can be searched very quickly indeed! This particular search method, called a *binary search*, is a very important technique and is the basis for many related algorithms. We will be discussing it in more detail in Chapter A3.

Iteration. Iteration techniques are usually applicable to numerical problems. An example is the computation of a function such as sin(X). It can be shown that the value of sin(X) is given by the expression

$$\sin(X) = X - X^3/3! + X^5/5! - X^7/7! + \ldots$$

(where 5! means 5 × 4 × 3 × 2 × 1, or *factorial* 5). This does not lead to a direct algorithm, because it requires an infinite number of operations. However, for any desired degree of precision, it is sufficient to use only the first part of the infinite sequence. In particular, if we are content with a precision of $\pm 10^{-5}$ for all values of X between -1.0 and $+1.0$, it can be shown that we can use

$$\sin(X) \cong X - X^3/3! + X^5/5! - X^7/7!$$

This computation requires only a finite number of operations, and can now be coded directly. If more precision is needed, additional terms can be added. For example, the next term ($X^9/9!$) should be added if an accuracy of 10^{-7} is required for the same values of X. It can be shown for this example that the desired precision can be achieved by including all terms until a term is generated that is smaller than the error allowed, so a program can be written to *iterate* until the desired accuracy is obtained, as shown in Program A1.2.

Program A1.2 ***Compute sine by iteration***

```
SINE: program
    The sine of a number X is computed using a power series. Terms
    are added until the next term is less than ERROR.
    real SINE,X,ERROR,NEXT__TERM,I
        SINE←X
        I←4.0
        NEXT__TERM←−X↑3 / 6.0
        do while ABS(NEXT__TERM)≥ERROR
            SINE←SINE+NEXT__TERM
            NEXT__TERM←−NEXT__TERM*X↑2 / (I*(I+1.0))
            I←I+2.0
            enddo
        output 'SINE OF',X,'IS',SINE
    endprogram SINE
```

Trial and error. In the trial-and-error form of iteration, the amount by which the current approximation fails to satisfy the problem is used to determine the next approximation. The square-root example in Chapter G3 is of this type. The current approximation X to the square root of 2 is squared and compared to 2. If it is less, the answer is tentatively increased by a small amount to try and get closer.

A trial-and-error process can be likened to the way people perform many everyday actions—driving a car, putting an object down, or almost any action involving movement. A person steers a car in the desired direction by turning the wheel approximately the correct amount and observing whether more or less turn is needed; that is, she observes the error in a trial attempt and then corrects to reduce the error. No matter how well the driver knew the route, a car could not be driven blindfolded, even if there were no other cars on the road, because the measurement of the error is essential to the correction. Chapter A2 gives a trial-and-error method for solving numerical problems common to many scientific and business applications.

Chapter

A2

The Method of Bisection

Suppose you have decided to buy a car costing \$3000, and the dealer says you can have it if you will put \$500 down and pay \$115 a month for 24 months. You will naturally ask yourself whether you can't find less expensive financing somewhere else. One approach is to call a number of banks and find out what interest rate they will charge on a loan for the balance of \$2500, to be paid monthly over 24 months. Question: What is the equivalent interest rate of the financing offered by the dealer? (Legally the dealer must tell you, but since the finance charges probably also include a number of other items, such as insurance against the buyer defaulting, the declared interest rate may not be the same as the effective interest rate.) If you could calculate the effective interest rate, you could make a quick decision whether to choose alternate financing from a bank.

If the original cost is COST (\$2500 in our example, because the amount to be borrowed is the balance after the down payment), the repayment rate is R per month (\$115 in our example), and the interest rate is P percent per year, then the amount left to be repaid after N months is

$$\mathsf{COST} \times \left(1 + \frac{\mathsf{P}}{1200}\right)^{\mathsf{N}} - \mathsf{R} \times \frac{\left(1 + \frac{\mathsf{P}}{1200}\right)^{\mathsf{N}} - 1}{\frac{\mathsf{P}}{1200}}$$

We would like to know what value of P makes this value zero when COST = 2500, R = 115, and N = 24; that is, we want to solve the equation

$$2500 \times \left(1 + \frac{P}{1200}\right)^{24} - 115 \times \frac{\left(1 + \frac{P}{1200}\right)^{24} - 1}{\frac{P}{1200}} = 0$$

for P.

The *method of bisection* is a technique for solving problems of this type. It is, in fact, one of the simplest and most reliable methods for finding the solution to an equation, though not one of the fastest. The solution of equations arises in almost all applications of computers. For example, an engineer who wants to find values of variables to achieve certain objectives—such as selecting the thrust of a rocket to place a spacecraft in the correct orbit—must usually solve an equation.

We can write the equation to be solved for X as

$$F(X) = 0$$

(In the financing problem, X is replaced by P, the interest rate.) Let us suppose that we can determine easily that the function F(X) changes sign as X changes from one value to another. For example, a quick calculation of the remaining balance after 24 months in the example above reveals that the balance is negative if the interest rate is 1% and positive if the interest rate is 50%. Therefore, we know that an interest rate somewhere between 1% and 50% will make the balance exactly zero.

In the method of bisection, we start with a function F(X) and two values of X for which the values of F(X) have opposite signs. This tells us that there is a value of X, between the two values given, for which F(X) is zero. (Mathematically, we are assuming that F(X) is *continuous*, as it is for most reasonable problems.) Suppose, for example, that the function F(X) is given by $X^3 + 3X - 5$. Figure A2.1 shows the graph of this function. As can be seen, the function is below the X-axis at $X = 0$ and above it at $X = 2$. Since the function is continuous, it must cross the

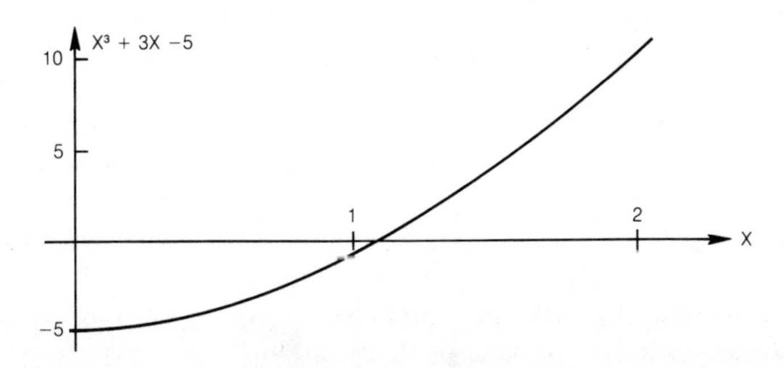

Figure A2.1 $X^2 + 3X - 5$

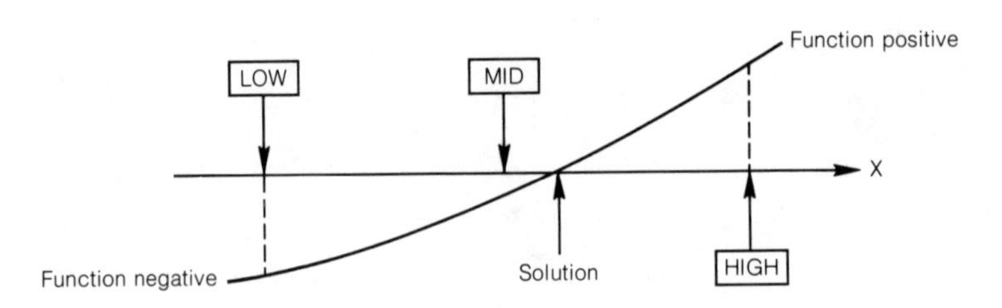

Figure A2.2 Relation of computed points to solution

X-axis somewhere between X = 0 and X = 2. Hence, for some value of X between 0 and 2, $X^3 + 3X - 5 = 0$.

If we set LOW to 0 and HIGH to 2, then there is a solution of the problem between LOW and HIGH, as shown in Figure A2.2. The important characteristic of the two points LOW and HIGH is that the value of the function is negative at LOW and positive at HIGH. Now let us find the point MID midway between LOW and HIGH, as shown in Figure A2.2. If we look at the sign of the function at MID we have two cases:

Case 1.

The function is negative at MID (as shown). Since the function is positive at HIGH, a solution lies between MID and HIGH. Consequently, we can move LOW over to the point MID to get Figure A2.3. (Note that the solution still lies between LOW and HIGH.)

Case 2.

The function is positive at MID. In this case a solution lies between LOW and MID. Consequently we can move HIGH over to the point MID, as shown in Figure A2.4.

In either case, we finish up with the points LOW and HIGH such that the function is still negative at LOW and positive at HIGH. Furthermore, the distance between LOW and HIGH is now half what it was initially.

This process can be repeated by setting MID to the midpoint between the new LOW and HIGH and repeating the analysis, as shown in Figure A2.5.

It can be seen that the points LOW and HIGH are getting closer to each other at each step. Since the solution of the original problem lies between LOW and HIGH after each step, we are getting a better and better approximation to the position of the solution X.

In the computer, we can represent only a finite number of values exactly. In general, the solution X will not be one of these values, so we

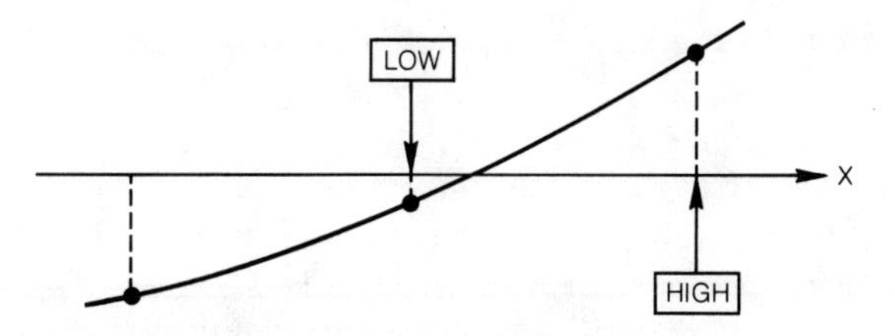

Figure A2.3 Halving the interval: negative at MID

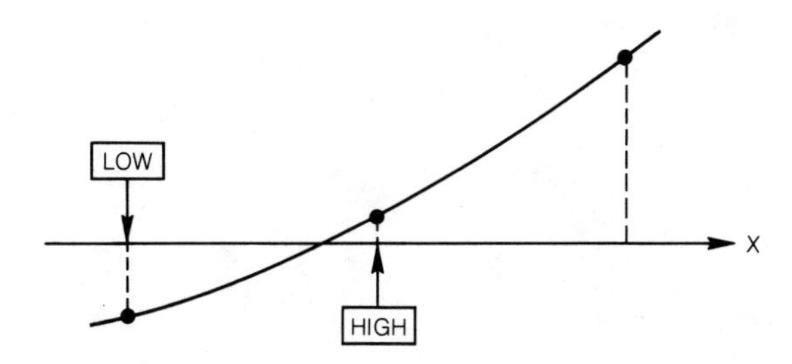

Figure A2.4 Halving the interval: positive at MID

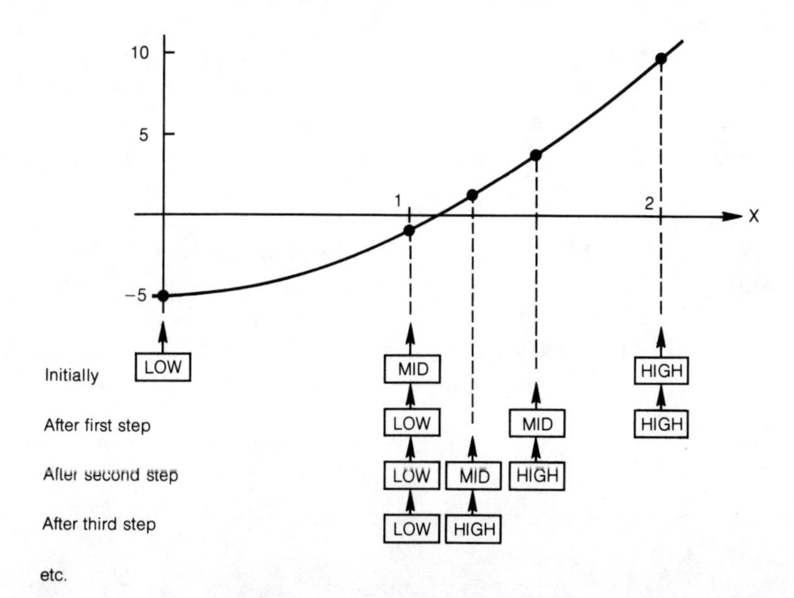

Figure A2.5 Successive approximations

Program A2.1 ***Solve equation by bisection***

```
BISECTION: program
   Program to solve the equation F(X) = 0 by the method of bi-
   section. Initial values assumed given for LOW and HIGH such
   that LOW_INIT < HIGH_INIT, F(LOW_INIT) < 0, and
   F(HIGH_INIT) > 0. Code to calculate value of function for any
   argument value X is also assumed given.
   real F,X,LOW,HIGH,LOW_INIT,HIGH_INIT
      LOW←LOW_INIT
      HIGH←HIGH_INIT
      do while (HIGH−LOW)>0.002
         X←(LOW+HIGH) / 2.
         F←value of function for argument X
         if F>0
            then HIGH←X
            else LOW←X
            endif
         enddo
      output (LOW+HIGH) / 2.
   endprogram BISECTION
```

can only hope to compute an approximation to it. Therefore we can stop when LOW and HIGH are within the precision we want. For example, if we want an answer accurate to 0.001, we can stop when HIGH − LOW ≤ 0.002. The solution can be taken to be the midpoint of LOW and HIGH. Since this is within 0.001 of both LOW and HIGH, it must be within 0.001 of the solution.*

Program A2.1 implements the procedure just described for finding the zero of an arbitrary function F, which is assumed to be specified somewhere. Initial values of LOW and HIGH such that LOW<HIGH, F(LOW)<0 and F(HIGH)>0, are also assumed to be given. (In practice, it would be necessary to search for such values first if they were not known.) The program uses the variable X instead of MID.

The if statement asks whether F is greater than zero. The case where F is exactly zero is lumped with the case where F is less than zero. The program could test specifically for the case F = 0, but the chance that the floating-point value of F is exactly zero is very small, so such a test would only waste time without reducing very much the average number of passes through the loop needed to find an approximation to a given degree of precision.

*Rounding errors can make this statement false.

Problems

1. Suppose we know only that the signs of F at X=LOW and X=HIGH are different. Modify Program A2.1 so that it works even if it is the sign of F(LOW) that is positive.
2. Use the method of bisection to solve the interest-rate problem discussed at the beginning of this chapter.
3. If we start the method of bisection with HIGH − LOW = 1.0, how many passes through the loop are necessary to achieve a precision of at least 0.001? What is the precision after N passes through the loop?

Chapter

A3

Searching and Sorting

Searching and sorting are important in many forms of data processing. Many applications work with large sets of data that must be kept up to date—bank records, student records, industrial inventories, store accounts, and so forth. In each of these applications the data is in the form of a list that must be accessed by means of a *key*: for a bank record, the key is probably the account number, for a student the student ID number, for an inventory the name of the item or its identification code, and for a store account the customer charge number. The data really consists of several related lists: for student records, these would include the student ID number, the student name, course records, grades, and other data. If the student ID number is known, we will probably want to use it as the key to access the lists; if the ID number is not known, we may want to use the student name as the key.

Let us consider as an example a list of student ID numbers in an array ID and a list of student scores in an array SCORE. Suppose we already have a set of N pairs of values in these arrays, and we wish to read a replacement score for one of the students. We will input the student ID number, say IDN, and a corresponding score, say SCN, and change the score corresponding to IDN to contain SCN. As we have pointed out earlier, if we have no knowledge of the order in which the ID numbers are stored, we will have to use a sequential search such as that shown in Program A3.1.

This program always takes N passes through the loop to find the ID number. It could be improved by stopping the loop when the desired ID is found, which would reduce the average number of passes through the loop to N/2 if the desired ID is equally likely to appear in any posi-

Program A3.1 ***Sequential search***

```
SEQ__SEARCH: program
    integer I,K,N,ID(N),SCORE(N)
        input IDN,SCN
        I←1
        do while I≤N
            if ID(I)=IDN then K←I endif
            I←I+1
            enddo
        SCORE(K)←SCN
    endprogram SEQ__SEARCH
```

Program A3.1a ***Improved sequential search***

```
SEQ__SEARCH: program
    integer I,N,ID(N),SCORE(N)
        input IDN,SCN
        I←1
        do while I≤N and ID(I)≠IDN
            I←I+1
            enddo
        if I≠N+1 then SCORE(I)←SCN endif
    endprogram SEQ__SEARCH
```

tion. At the same time, we should allow for the possibility that the input data is incorrect and IDN does not appear in the list at all. The revised program appears in Program A3.1a. The loop continues as long as both the condition I≤N and the condition IDN≠ID(I) remain true. If IDN is not present, it still takes N passes; otherwise it is faster.

A3.1 BINARY SEARCH

To develop a faster search method, we must know something about the way the data is organized. If we assume that the array ID is in ascending numerical order, then other types of search can be used. The fastest search for large arrays is the *binary search* mentioned in Chapter A1. Program A3.2 gives a version of binary search. It uses the two integer variables LOW and HIGH to hold the lowest and the highest index of a block of cells within the array ID that must contain IDN if it is present at all. Initially, LOW is set to 1, the beginning of the array, and HIGH is set to N, the end of the array. Each pass of the loop finds the index of the middle entry in the "active" part of the array by computing (LOW+

Program A3.2 ***Binary search in an ordered array***

```
BIN__SEARCH: program
    This program reads a student ID number IDN and prints the corre-
    sponding score. Contents of the arrays ID and SCORE and their
    dimension N are assumed given. If the requested ID number IDN
    is not found in the array ID, an error message is printed.
    integer I,N,ID(N),SCORE(N),IDN,LOW,MID,HIGH
        input IDN
        I←0
        LOW←1
        HIGH←N
        do while LOW≤HIGH and I=0
            | MID←(LOW+HIGH)÷2
            | if ID(MID)=IDN
            |     | then I←MID
            |     | else if ID(MID)<IDN
            |     |     | then LOW←MID+1
            |     |     | else HIGH←MID−1
            |     |     | endif
            |     | endif
            | enddo
        if I≠0
            then output 'SCORE FOR STUDENT',IDN,'IS',SCORE(I)
            else output 'NO DATA FOR STUDENT',IDN
            endif
    endprogram BIN__SEARCH
```

HIGH)÷2 (this will be the midpoint if LOW+HIGH is even; otherwise, it will be the lower of the two entries nearest the midpoint). If this entry is the one sought, there is nothing else to do, so the variable I is set nonzero to flag the end of the search. Otherwise, either LOW is raised to MID+1 or HIGH is lowered to MID−1, to reduce the size of the active segment by a factor of two. The choice is made by comparing the entry at MID with the value sought, as illustrated in Figure A3.1. The loop continues until either the desired entry is found or there is no array left to search. The latter condition is detected when LOW>HIGH. We can guarantee that one of the terminating conditions for the loop will eventually become true because each pass either sets I nonzero (which terminates the loop at the start of the next pass), moves LOW up by at least one, or moves HIGH down by at least one, until LOW>HIGH. (We can make the last statement from the observation that as long as LOW≤HIGH, then LOW≤(LOW+HIGH)÷2≤HIGH.)

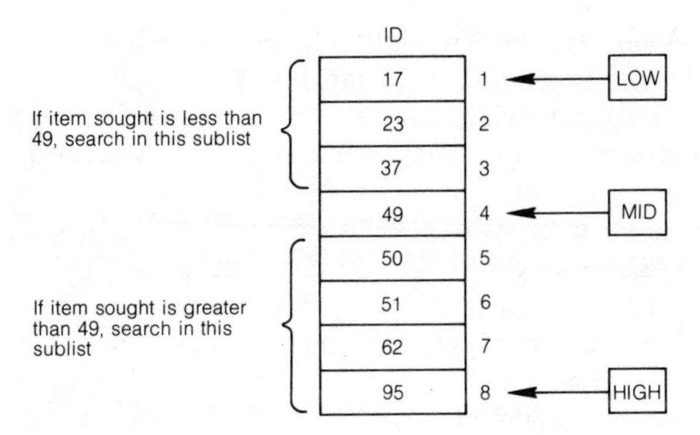

Figure A3.1 Binary search in array ID

It is instructive to follow the flow of the binary-search program for several cases. Suppose the array ID contains the entries 17, 23, 37, 49, 50, 51, 62, and 95 in position ID(1) through ID(8). Table A3.1 shows a search for IDN = 23. Each row in the table shows the states of the variables right after the statement assigning a value to MID has been executed. The last column describes the action that will be taken in the rest of the loop. As can be seen, the value 23 is located with just two passes through the loop. In Table A3.2, the ID number 61 is sought. It is not present, but only three passes through the loop are needed to determine this fact, compared with eight passes that would be needed with a sequential search.

The true power of the binary-search method is realized in very large tables of data, because for each doubling of the size of the table, the

TABLE A3.1 EXAMPLE OF BINARY SEARCH

IDN	LOW	HIGH	MID	ID(MID)	Action
23	1	8	4	49	assign 3 (= MID − 1) to HIGH
23	1	3	2	23	assign 2 (= MID) to K
Loop terminates because K ≠ 0					

TABLE A3.2 EXAMPLE OF BINARY SEARCH

IDN	LOW	HIGH	MID	ID(MID)	Action
61	1	8	4	49	assign 5 (= MID + 1) to LOW
61	5	8	6	51	assign 7 (= MID + 1) to LOW
61	7	8	7	62	assign 6 (= MID − 1) to HIGH
Loop terminates because LOW > HIGH					

search takes only one more pass in the worst case. Although the program itself is more complex than the program for a sequential search, increasing the table size will eventually make the binary search faster. For example, suppose a sequential search in a table of 15 items takes a maximum of 15 microseconds, and a binary search in the same size table takes a maximum of 60 microseconds. For a table of 120 items, a sequential search will take a maximum of 120 microseconds. A little calculation shows that a binary search takes at most four passes for 15 items, so that each pass is taking 15 microseconds. Therefore, double the number of items (30) can be searched in 75 microseconds, double again (60) in 90 microseconds, and double again (120) in 105 microseconds, 15 microseconds faster than the sequential search. The advantage really starts to show up at the next doubling to 240 items, which takes 120 microseconds for the binary search versus 240 for the sequential search, assuming our original timing figures are correct. (In fact, this example is biased in favor of the sequential search—the binary search will generally be faster for tables of more than about 20 items.)

A3.2 SORTING

We have seen that it is possible to use faster search methods if the data is sorted. Sometimes the data we have to work with is naturally sorted. We may, for example, be working with data read from a master file, such as a customer list. If the master file is sorted, then any data we extract from it will be sorted. In that case, the choice of a search method, such as sequential or binary, is based simply on the expected number of items in the list. In other cases we have a list of data that is unordered, and we must decide whether to sort it for the sake of a faster search or to keep it unordered. Sorting takes a lot longer than searching for a few items, so it is not normally worthwhile unless many references will be made to the data or there is some other reason to sort it. (If we want to print the data in sorted order later, for example, we may as well sort it when we first get it and take advantage of the ordering in the rest of the job.)

In this section we will look at three simple sort programs, each a little better than the one before. It is possible to write even faster programs, using methods that are too complex to discuss here. The sorting methods presented here have the advantage of being simple, and therefore easy to code and test. Furthermore, for small amounts of data they are as efficient as more complex algorithms that use fewer passes through loops but take more computer time in each loop.

The first method we will examine is called the *selection sort.* If we want to sort an N-element array into ascending order, we first find the

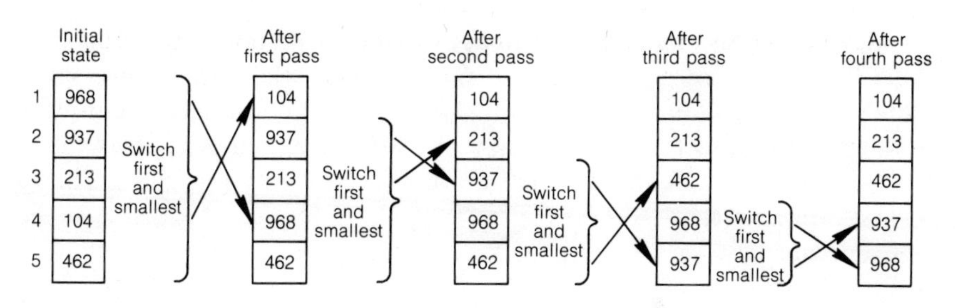

Figure A3.2 Stages of selection sort

smallest entry and move it into the first position, by switching it with the element originally in that position. Then we know that the first element is correct, and we can proceed in the same way to sort the remaining N − 1 elements in positions 2 through N of the array. This process is shown in Figure A3.2 for N = 5. Each successive pass moves one more element into position. After N − 1 passes, the array will be in order. The overall form of the algorithm is shown in Program A3.3. Each pass through the outer loop moves the smallest element of the array ID(I), for I from K to N, into position ID(K). All that remains is to "refine" the two steps of finding the largest value in the subarray and switching the elements. Those sections of code are shown in the refined version, Program A3.3a.

The sorting algorithm of Program A3.3a requires a long time to execute. On the first pass through the outer loop, the inner loop is executed for I from 2 to N; on the second pass, from 3 to N; and so on. The total number of passes needed through the inner loop is thus (N − 1) +

Program A3.3 ***Selection sort***

```
SELECT__SORT: program
    Program to sort an integer array into ascending order using a selec-
    tion sort. The contents of the array ID and its dimension N are
    assumed given.
    integer K,N,ID(N)
        K←1
        do while K≤N−1
            First K − 1 elements are now in order. Find smallest remain-
            ing element and move it to position K.
            set I to index of smallest element between ID(K) and ID(N)
            switch ID(I) and ID(K)
            K←K+1
            enddo
    endprogram SELECT__SORT
```

Program A3.3a ***Further refinement of Program A3.3***

```
SELECT__SORT: program
    Program to sort an integer array into ascending order using a selec-
    tion sort. The contents of the array ID and its dimension N are
    assumed given.
    integer I,J,K,N,ID(N),SMALL
        K←1
        do while K≤N−1
          | First K − 1 elements are now in order. Find smallest remain-
          | ing element and move it to position K.
          | I←K
          | SMALL←ID(K)
          | J←K+1
          | do while J≤N
          |   | SMALL contains smallest element between ID(K) and
          |   | ID(J−1). I is its index.
          |   | if ID(J)<SMALL
          |   |   | then I←J
          |   |   |      SMALL←ID(J)
          |   |   | endif
          |   | J←J+1
          |   | enddo
          | ID(I)←ID(K)
          | ID(K)←SMALL
          | K←K+1
          | enddo
    endprogram SELECT__SORT
```

(N − 2) + (N − 3) + . . . + 3 + 2 + 1, which comes to N × (N − 1)/2.* If N is 100, this means the inner loop must be executed 4950 times, regardless of the original order of the data: even if it is already in order and needs no sorting, 4950 passes through the inner loop are still required.

The observation that the sorting method just described can take a long time even if the elements are already in order prompts us to look

---

*This can be seen by writing the series once in the order given and once in reverse:

$$\begin{array}{ccccccccccccc} (N-1) & + & (N-2) & + & (N-3) & + & \ldots & + & 3 & + & 2 & + & 1 \\ 1 & + & 2 & + & 3 & + & \ldots & + & (N-3) & + & (N-2) & + & (N-1) \end{array}$$

Adding term by term, we find that twice the sum of the series is

$$N + N + N + \ldots + N + N + N$$

Since there are N − 1 terms, twice the sum of the series is N × (N − 1), so the sum of the series is N × (N − 1)/2.

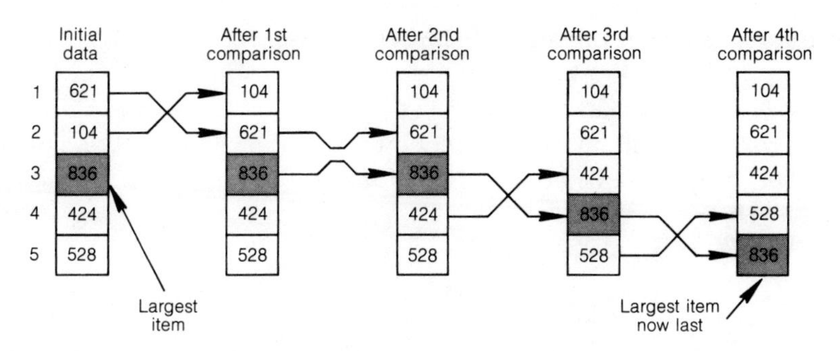

Figure A3.3 First pass of bubble sort

for methods that are faster when the data is already sorted or nearly so. The next method we examine is called the *bubble sort.* On each pass through the list, each pair of adjacent items is compared: if they are in the proper order they are left alone and the next pair examined; if they are out of order they are switched before going on to the next pair. The process is illustrated in Figure A3.3, which shows the following steps:

1. Compare the first pair, 621 and 104. They are out of order, so switch them.
2. Compare the second pair, 621 and 836. They are already in order, so leave them alone.
3. Compare the third pair, 836 and 424. They are out of order, so switch them.
4. Compare the fourth pair, 836 and 528. They are out of order, so switch them.

As a result of these operations, at the end of the first pass the largest element in the list, 836, has "bubbled" to the last position. The next pass will work only with the remaining elements, and will move the next largest to the next-to-last position, and so on. Notice that if an N-element list is already sorted to begin with, we will find this out on the first pass, after only N − 1 comparisons—the minimum possible to determine that N elements are already sorted.

The bubble-sort algorithm is shown in Program A3.4. I is the loop counter, and is decremented from N to 2 on successive passes through the outer loop. On each pass, the counter for the inner loop, J, need run only from 1 to I − 1, since the last N−I positions already contain the N−I largest elements. At most N − 1 passes through the outer loop are needed to sort the list into ascending order, but it may take less: if no switches are made on any given pass, then we are finished. Accordingly, a variable K is used to remember whether any switches have yet been

Program A3.4 ***Bubble sort***

```
BUBBLE_SORT: program
   Program to sort an integer array into ascending order using a bubble
   sort. The contents of the array ID and its dimension N are assumed
   given.
   integer I,J,K,N,ID(N), TEMP
      I←N
      K←1
      The flag K is nonzero on the first pass of the outer loop and
      whenever switches were made on the previous pass.
      do while I≥2 and K≠0
        | The largest N − I elements of ID are now in order in posi-
        | tions ID(I+1) to ID(N). Float the largest of the first I elements
        | to position ID(I).
        | J←1
        | K←0
        | do while J≤I−1
        |   | The largest of the first J elements is now in position
        |   | ID(J).
        |   | if ID(J)>ID(J+1)
        |   |   | then
        |   |   |      Switch entries ID(J) and ID(J+1).
        |   |   |      TEMP←ID(J)
        |   |   |      ID(J)←ID(J+1)
        |   |   |      ID(J+1)←TEMP
        |   |   |      K←1
        |   |   | endif
        |   | J←J+1
        |   | enddo
        | I←I−1
        | enddo
   endprogram BUBBLE_SORT
```

made on the current pass: it is set to zero at the beginning of the pass and changed to nonzero whenever a switch is made. If K is still zero at the beginning of the next pass, the outer loop terminates; otherwise it terminates when I becomes less than 2.

In the best case, when the data is already in order, the bubble sort takes only one scan through the list, and no sorting method can do better. However, in the worst case, the bubble sort takes a maximum of $N \times (N - 1)/2$ comparisons, as does any sequential method. (The worst case occurs when the data is in reverse order: a bubble sort of 5, 4, 3, 2, 1 into ascending order requires 10 switches. After the first pass

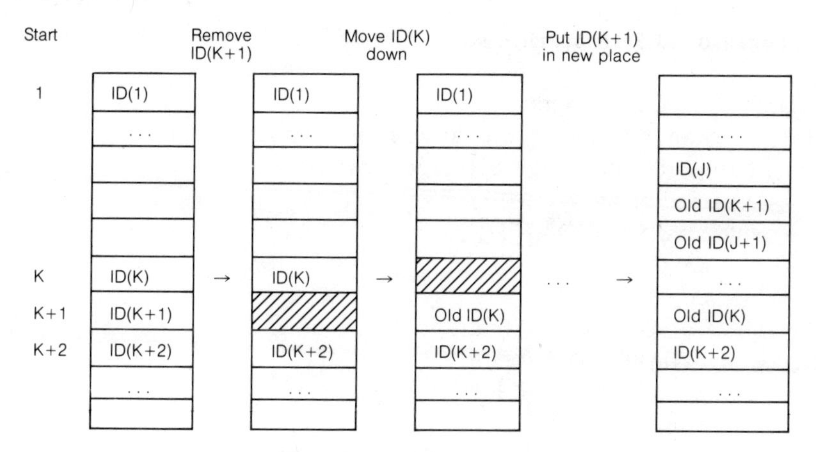

Figure A3.4 Insertion sort

through the outer loop, four switches have been made and the list is 4, 3, 2, 1, 5.) Furthermore, in the worst case, each of these comparisons will cause a switch of two data items (which requires three assignment statements) compared with the maximum of N − 1 switches needed in the selection sort described earlier. Thus, although the bubble sort is faster than the earlier method in the best case, it is slower in the worst case.

The third sorting method we will consider is the *insertion sort*, illustrated in Figure A3.4. In this method, the list is scanned until an out-of-order element is found. The scan is then temporarily halted while a backward scan is made to find the point at which to insert the out-of-order element. Elements bypassed during this backward scan are moved up one position to make room for the element being inserted. Thus, in Figure A3.4, if the first K elements are in order but the K + 1st is not—that is, if ID(1)≤ID(2)≤ID(3)≤ . . . ≤ID(K) but ID(K)>ID(K+1)—then ID(K+1) is out of order and must be moved to an earlier position. It is removed from the array (that is, a copy of its value is saved somewhere, say at TEMP) and the elements ID(K), ID(K−1), . . . are examined until an ID(J) is found such that ID(J) ≤ old ID(K+1). In the process, the old elements ID(J+1) through ID(K) will have been moved to positions ID(J+2) through ID(K+1), clearing the J + 1st position for the value saved at TEMP.

Program A3.5 is based on this method. The inner loop is in the **then** clause of an **if** statement, and is executed only when an out-of-order element is found. Like the bubble sort, the insertion sort is a sequential sort technique. Like the bubble sort, it requires only N − 1 comparisons in the best case (when the data is already sorted initially); again like the bubble sort, it requires N × (N − 1)/2 comparisons and moves in

Program A3.5 *Insertion sort*

```
INSERT__SORT: program
   Program to sort an integer array into ascending order using an inser-
   tion sort. The contents of the array ID and its dimension N are
   assumed given.
   integer J,K,N,ID(N),TEMP
      K←1
      do while K≤N−1
         | ID(1) to ID(K) are now in ascending order. Extend to
         | ID(K+1).
         | if ID(K)>ID(K+1)
         |    | then
         |    |    Move ID(K+1) back to its place in array. First save
         |    |    its value in TEMP, then move earlier elements up
         |    |    one place until the right position is found.
         |    |    TEMP←ID(K+1)
         |    |    J←K
         |    |    do while J≥1 and ID(J)>TEMP
         |    |       | ID(J+1)←ID(J)
         |    |       | J←J−1
         |    |       | enddo
         |    |    ID(J+1)←TEMP
         |    | endif
         | K←K+1
         | enddo
   endprogram INSERT__SORT
```

the worst case—but unlike the bubble sort, it completes each move in only one assignment statement instead of three.

Any method of sorting based on a sequential search through an N-element list for the largest remaining element will take time proportional to N^2, as do the three methods discussed so far. This means that the execution time increases rapidly as N gets large. We saw earlier that the sequential search becomes progressively worse compared to the binary search as N gets large. There are, in fact, sorting methods that employ techniques similar to those used in a binary search, and are faster than sequential methods for large amounts of data. The *binary merge*, for example, is a divide-and-conquer algorithm that is useful for large files. Figure A3.5 illustrates the method. The file is divided into two subfiles of about equal length, and each subfile is sorted by some method. The two subfiles are *merged* into one file by repeatedly selecting the next item from the file with the smallest item. In the figure, the original file contains the values 7, 4, 9, 12, 1, 19, 5, and 21; it is divided into two sub-

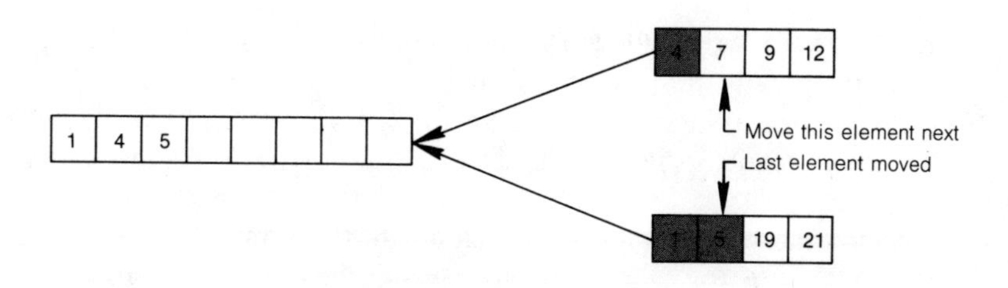

Figure A3.5 Binary merge

files, containing 7, 4, 9, and 12, and 1, 19, 5, and 21, respectively. Sorting each subfile separately, we get 4, 7, 9, and 12, and 1, 5, 19, and 21. These can now be merged to get 1, 4, 5, 7, 9, 12, 19, and 21. How do we sort the subfiles? If they are not too long, we can use one of the methods discussed earlier; otherwise, we can treat them the same way as the original file by subdividing them further. If, for example, we start with a file of 1024 items, we can subdivide it several times to get 128 files of 8 items each. Each of these subfiles can be sorted with a bubble sort, taking a maximum of $128 \times (8 \times (8 - 1)/2)$ comparisons. The sorted files can then be merged pairwise to get 64 files of 16 items each, taking another 1024 comparisons. Six merges later, we will have one file of 1024 items, at a total cost of $7 \times 1024 + 128 \times (8 \times (8 - 1)/2)$, or 10,752, comparisons. In contrast, the bubble sort can take as many as $1024 \times (1024 - 1)/2$, or 523,776, comparisons—more than 50 times as many.

Merging is the preferred sorting method for very large files, such as those that arise in many business data-processing applications. In such applications, the *master file* is already in order, and update information must be sorted and merged into it.

Problems

1. What is the maximum number of items that can be searched with a binary search if no more than
   a. 3
   b. 4
   c. 6
   d. n

   comparisons can be used in the worst case? (Assume that one comparison will determine whether one value is smaller than, equal to, or larger than another.)
2. If a single pass through a sequential search loop takes 2 microseconds and a single pass through a binary search loop takes 21 microseconds, construct a graph showing the worst-case search time for each

method versus the number of items in an array. For what number of items N will the binary search method be faster?

***3.** a. What arrangement of the integers 1, 2, 3, 4, and 5 in a five-element array will lead to the worst-case sort time for the insertion sort in Program A3.5?

b. How many assignments are made into the array ID in the worst case if N = 6 for each of the three methods of sorting discussed? (Remember that the worst cases may be different for each method.)

c. Can you generalize this result for any N?

4. If you are requested to write a program that will merge several (perhaps a hundred) new items into a very large existing file of information that is already sorted, which of the following two methods would you use, and why?

(i) Search through the sorted file to find the position of one of the new items, and insert it. (Don't worry about how to insert an item; assume it can be done.) Process each of the new items the same way.

(ii) First sort the new items and then merge the two sorted files: scan through the large file to find where the first entry from the small file of new entries will fit, and continue scanning to locate the position for each entry in the new file.

(In large business applications, the existing sorted file, which may be the master file of, say, bank customers, is kept on a secondary-storage device such as a disk file (discussed in Chapter C4). To scan the file and insert new information means reading the file into main memory a piece at a time, updating that piece, and writing it back out. It can take a long time to scan the whole file.)

Chapter A4 Pointers

The sorting techniques discussed in Chapter A3 all required moving data around to put it in the right order. If each data item is large (occupies a large number of memory cells), this movement can absorb large amounts of computer time. Suppose, for example, that we have two arrays: ID(I) contains the ID number of the Ith student, and SCORE(I,J) contains the Ith student's score on the Jth exam. If we use any of the sorting methods presented in Chapter A3 to sort the data by student ID number before printing, we will have to swap two whole rows in the array SCORE every time we swap a pair of elements in ID. If we don't, we will lose track of the correspondence between elements of ID and rows of SCORE.

We can overcome this difficulty by using an auxiliary array of *pointers* to ID and SCORE. (Some languages provide special pointer-valued variables and pointer-manipulation operations for this purpose, but to keep things simple we will use array indices instead. The basic principles are the same.) Thus we will use an integer array INDEX whose elements are indices into the arrays ID and SCORE. When we sort the student records into ascending order by ID number, we can simply move the indices in INDEX instead of moving the data itself in ID and SCORE. The method is depicted in Figure A4.1. The first three columns show the state of the arrays when the data is first entered into the computer; the last three show the arrays after the sort is completed. Notice that only the elements of the array INDEX have been moved: ID and SCORE remain unchanged. To print the student data in order by ID number, we simply step through the array INDEX, using each successive element as an index to help us find the entries to be printed next in ID and SCORE.

Pointers are a means for accessing data in memory without searching for it, and are one of the most important ideas in programming. They

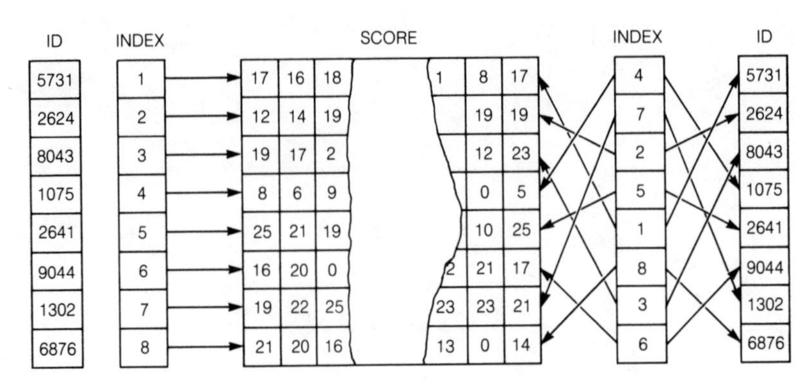

Figure A4.1 Sorting using pointers in INDEX

can be used in a variety of ways, in both business and scientific applications, to make our programs more efficient. In those applications to which they are suited, they can also make a program conceptually cleaner and easier to read.

Suppose we are writing an airline reservation system and want to keep a list of the names of the passengers on each flight. There is a separate such list for each flight, and each of these lists is constantly changing as reservations are made and canceled. (We will not concern ourselves in this discussion with the way character strings are handled—we will simply assume that we can use an array NAME to store the passenger names for a flight. We will also assume we know the maximum size such a list can assume—presumably the number of seats on the plane—so that we can allocate the right size array.)

Initially our array NAME is empty. As reservations are made on the flight, each new name can be added in the next available empty location in the array. To keep track of this location we will use a variable, say AVAIL, which is initially set to 1 (pointing to the first element of the array). To add a new name to the list, we store it at NAME(AVAIL) and increase AVAIL by one. Deleting an entry causes problems, however, because it leaves a "hole" in the array. We would like to use this space for another passenger, so we must either keep track of the holes or move the existing data around, as in Figure A4.2, to fill the holes as they are created. The latter alternative can be quite time-consuming, since half the entries (on the average) must be moved every time an entry is deleted. If the data is ordered, entries must be moved in this way to do insertions as well as deletions.

Once again, pointers provide the solution. This time we use them to construct a data structure called a *chained* or *linked list*, in which each entry carries a pointer to the next (see Figure A4.3). To scan the list in order, we begin with variable START, which points to the first

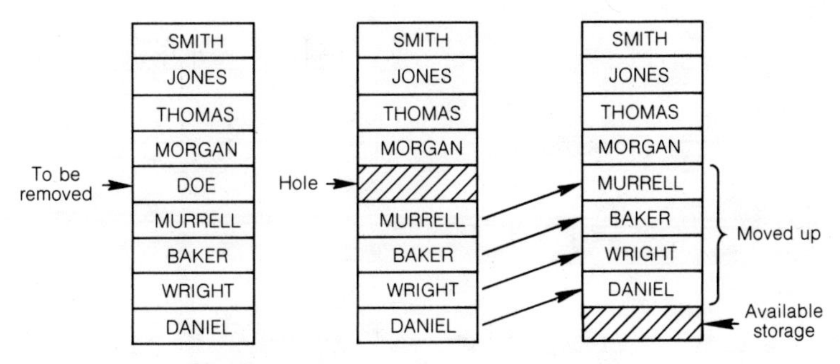

Figure A4.2 Deleting an entry in an array

item in the list; thereafter, each item we examine tells us where to find the next item. The end of the list is marked by a special "null" pointer, which can be represented by any recognizable value, such as zero. Using this data structure we can delete an item without leaving a "hole" in the list, simply by copying the pointer from that item into the item preceding it in the list, so that it "points around" the deleted item. In Figure A4.3, for example, we delete the name DOE from the list by copying its pointer (6) into the entry containing the name MORGAN, so that MORGAN points "around" DOE to MURRELL. To remove the last entry, DANIEL, we would copy its null pointer into the entry containing WRIGHT; to remove the first entry, SMITH, we would copy its pointer (2) into the variable START.

Notice, however, that in order to find the item preceding a given item, say DOE, in the list, we have to begin with the variable START and scan the list in order until we find an item pointing to DOE. Notice also that we cannot easily access a random element in the middle of

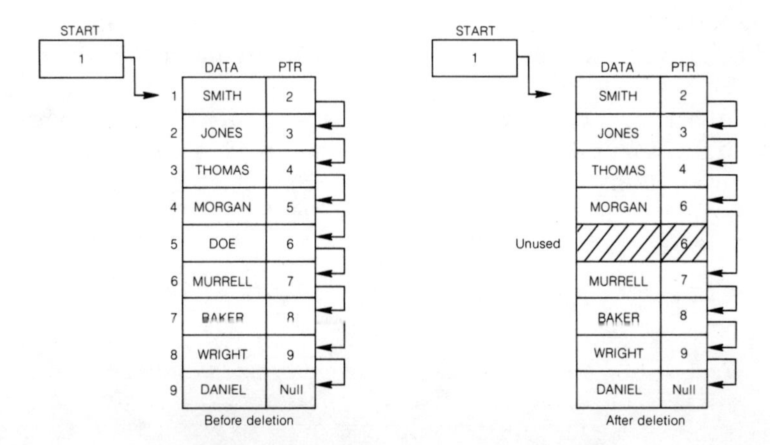

Figure A4.3 Information stored with pointers

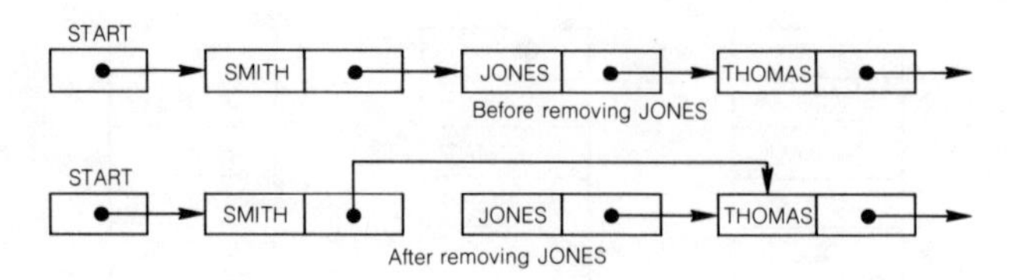

Figure A4.4 Removing an item from a list

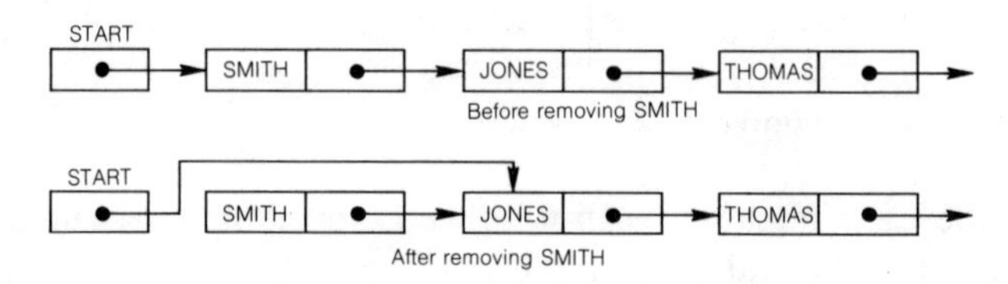

Figure A4.5 Removing the first item from a list

the list, since the I th element need not be stored in the I th location in the array. To find the I th element, we must again run through the list from the beginning, counting elements as we go until we reach the one we want. Thus the use of a chained list makes some operations (deleting an element, scanning the list in sequence) easier or more efficient and others (accessing the I th element) harder or less efficient. It also uses more memory space, since room must be allocated for the pointers as well as the data items themselves. The decision whether to use such a list depends on how often we expect to be performing the various operations.

Figures A4.4 and A4.5 depict the operations of deleting an element from the middle and the beginning of a list, respectively. From now on we will draw chained lists as shown in these figures. We do not know or care where the list items actually are in memory in relation to each other, because the pointers always tell us where the next item is.

When we add an item to a list, we can insert it anywhere we want in the chain of pointers connecting the list elements. If there is no reason to put it in any particular place, the front of the list is the easiest place to reach. Adding a new entry CAR to the front of a list containing MAP, TEMP, and B1 is shown in Figure A4.6. No data has to be moved: the

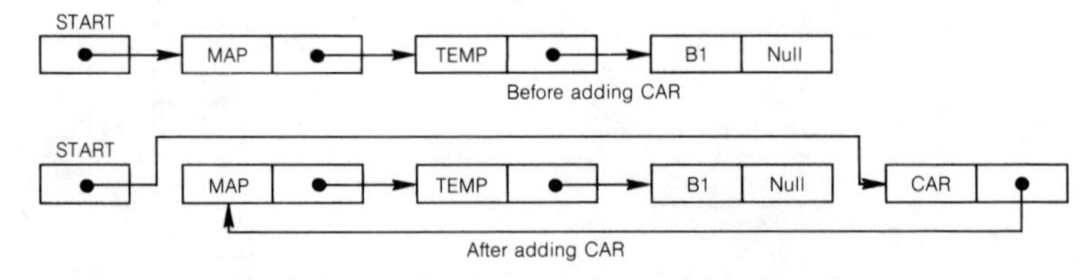

Figure A4.6 Inserting a new entry to the beginning of a list

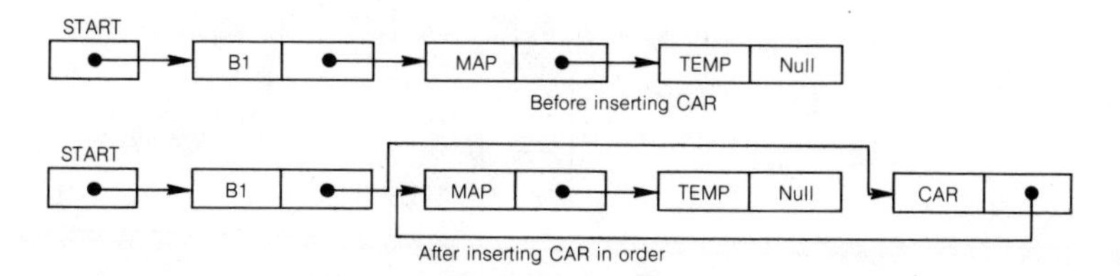

Figure A4.7 Inserting a new entry in alphabetical order

pointer that was in START is simply put into the new entry, and START is set to point to this new entry.

If a list is to be kept in sorted order, a sequential search is necessary to find the appropriate position for the new entry. Once the position is found, however, the new entry can be inserted without moving existing data: we need only set the pointer of the preceding entry to point to the new one, and set the pointer of the new one to point to the next one, as shown in Figure A4.7.

Each entry in Figures A4.4 through A4.7 is shown as a doublet. It could be a pair of adjacent memory cells or a pair of corresponding positions in two arrays: an array of character strings for the names and an array of integers for the pointers. We will call these arrays DATA and PTR respectively. These are shown side by side in Figure A4.8.

To add an entry to a chained list, we must find storage space for the new entry. The simplest method is sequential storage assignment: we initially assign an area of memory for the list and note that the first word in this area is currently available. As each entry is placed into the list, the first available locations in the assigned area are taken for the entry, as shown in Figure A4.8. AVAIL contains the address of the first available location in the storage area. When CAR is logically inserted at the front of the list containing MAP and B1, AVAIL and START are both updated appropriately.

When an entry is removed from the list, the storage it occupied is lost in this method unless it happens to be physically the last. Thus, if B1 is removed from the list in Figure A4.8, we get the list shown in

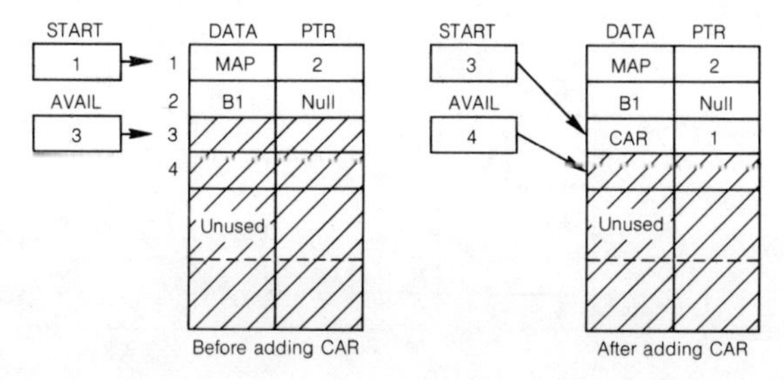

Figure A4.8 Available memory locations

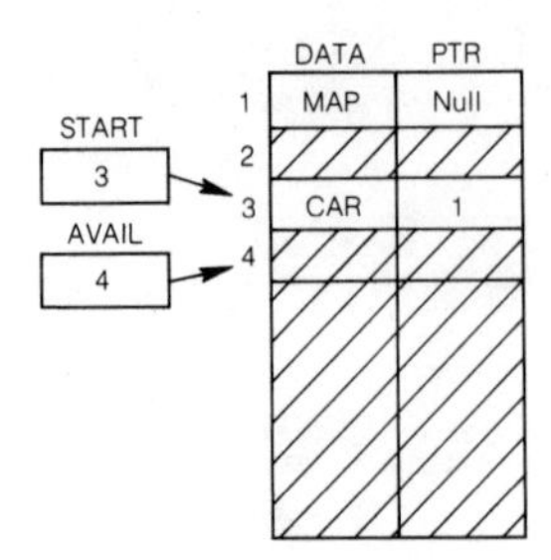

Figure A4.9 Lost storage

Figure A4.9. DATA(2) and PTR(2) are unused, but there is no simple way of reusing them for the next entry. We can solve this problem by keeping the unused space on a *free-storage list*, or *free list*. A variable, say FREE, can be used to hold a pointer to an unused element. That element can contain a pointer to another unused element, and so on. Initially, all available storage space must be put on the free list. This can be done by setting FREE to 1, PTR(I) to I + 1 for $1 \leq I \leq N - 1$, and PTR(N) to NULL. When an element is added to the list, space can be taken from the beginning of the free storage list; when an element is removed, the space released can be added to the free list. The steps needed to add CAR and then remove B1 are shown in Figure A4.10.

Programs A4.1 through A4.4 are subprograms to initialize the free list, insert a name in alphabetical order, delete a name, and output the list in order. One change has been made to the process described above: the pointer to the start of the data list is kept in PTR(1), which means that DATA(1) is not used. This avoids a number of complications in the special case of an empty list. (The reader can see this by rewriting Programs A4.2 and A4.3, using the variable START instead of PTR(1).) The pointer we are keeping in PTR(1) is sometimes called the *list head*.

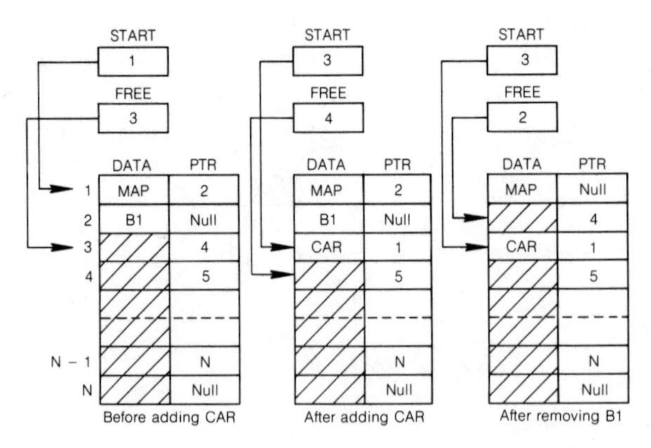

Figure A4.10 Use of free-storage pointer

Program A4.1 ***Initialize list pointers***

```
INIT__LIST: subprogram (PTR,N,AVAIL)
    integer I,N,PTR(N),AVAIL
    Initialize list head and start of free list.
        AVAIL←2
        PTR(1)←0
        Set pointers for free list.
        do for I←2 to N−1
            PTR(I)←I+1
            enddo
        Set null pointer at end of free list.
        PTR(N)←0
        return
    endsubprogram INIT__LIST
```

Program A4.2 ***Insert element in ordered list***

```
INSERT: subprogram (DATA,PTR,N,AVAIL,NAME)
    Program to insert the name NAME in the alphabetically ordered list
    contained in arrays DATA and PTR. Head of data list assumed to be
    in PTR(1), head of free list in AVAIL.
    integer I,J,N,PTR(N),AVAIL
    character(20)DATA(N),NAME
        if AVAIL=0
            then output 'NO FREE STORAGE LEFT'
            else
                I←1
                Search list for insertion point for NAME.
                do while PTR(I)≠0 and DATA(PTR(I))<NAME
                    I←PTR(I)
                    enddo
                I now contains entry of last element in list less than
                NAME. Allocate space from free list for new entry and
                insert it following entry I by setting pointers.
                J←AVAIL
                AVAIL←PTR(J)
                PTR(J)←PTR(I)
                DATA(J)←NAME
                PTR(I)←J
            endif
        return
    endsubprogram INSERT
```

Program A4.3 ***Delete element from chained list***

```
DELETE: subprogram (DATA,PTR,N,AVAIL,NAME)
    Program to delete the name NAME from the chained list contained
    in arrays DATA and PTR. If NAME is present, it is removed from
    data list and the space it occupies returned to free list; if not present,
    an error message is printed. Head of data list assumed to be in
    PTR(1), head of free list in AVAIL.
    integer I,J,N,PTR(N),AVAIL
    character(20)DATA(N),NAME
        I←1
        Find NAME in data list.
        do while PTR(I)≠0 and DATA(PTR(I))≠NAME
            I←PTR(I)
            enddo
        if PTR(I)=0
            then output 'NAME',NAME,'NOT PRESENT IN LIST'
            else
                Entry following I now contains NAME. Fix pointers to
                delete it from data list and put it on free list.
                J←PTR(I)
                PTR(I)←PTR(J)
                PTR(J)←AVAIL
                AVAIL←J
            endif
        return
    endsubprogram DELETE
```

Program A4.4 ***Print chained list in order***

```
PRINT__LIST: subprogram (DATA,PTR,N)
    Program to print the chained list contained in arrays DATA and
    PTR. List head assumed to be in PTR(1).
    integer I,N,PTR(N)
    character(20)DATA(N)
        I←PTR(1)
        do while I≠0
            output DATA(I)
            I←PTR(I)
            enddo
        return
    endsubprogram PRINT__LIST
```

Notice that the do while statements in Programs A4.2 and A4.3 may cause problems on some systems because of a reference to DATA(0). In this case, the do while statement in Program A4.2 must be replaced with the equivalent pair of statements

```
do while PTR(I)≠0
    if DATA(PTR(I))≥NAME then exit endif
```

and the one in Program A4.3 with the statements

```
do while PTR(I)≠0
    if DATA(PTR(I))=NAME then exit endif
```

Problems

1. Notice that Program A4.3 does not make use of the fact that the data is in alphabetical order. This means that if we try to delete a name that is not in the list, we will examine every element in the list looking for it. Change Program A4.3 so that it does not search past the first entry in the list larger than NAME. (This cuts the execution time in half when NAME is not present.)

2. *a. Write a subprogram to add a name to the end of an *unordered* list stored in the same way as in Programs A4.1 through A4.4. Your subprogram should take an additional parameter, END, containing the index of the last element in the list.

*b. Modify Program A4.3 to accept a similar parameter END and reset it appropriately if the last element in the list is deleted.

c. What value should be stored in END when the list is empty? (This value should be chosen to avoid having to test for special cases in the program.)

3. Write a program to sort an unordered chained list, using the bubble sort method given in Chapter A3.

4. Why is it not advisable to use a binary search on an ordered list?

***5.** An insertion sort cannot be performed on a chained list without some fancy footwork, because the insertion sort scans in both directions: forward to see if anything is out of order, and backward to move an out-of-order item to its proper place. Here is a technique that can be used:

Search forward, looking for an element that is out of place; at the same time, reverse the pointers so that there is a backward chain through the elements that have already been scanned. This chain can be used when it is necessary to move an element back. When the forward scan has been completed, it would appear that we have to scan back through the chain to reverse the pointers again. However, this can be avoided by sorting the list in the reverse order in the first place, so the chain is left in increasing order at the finish.

Write a program to perform such a sort. It should start with an unordered list in arrays DATA and PTR, with the list head in PTR(1), and finish with a list in increasing order.

6. Write a subprogram that takes two ordered lists, starting in LIST1 and LIST2 (that is, the first element of the first list is in DATA(LIST1) and PTR(LIST1)), and returns an ordered list containing the elements of both lists. The result should start in LIST3. No elements of the array DATA should be changed, only the pointers in PTR.

7. Write a version of Program A4.2 under the assumption that the start of the list is in a variable START (which should be an additional parameter). Note the additional code needed when a list head in the PTR array is not used.

Chapter

A5

Series Methods

Scientists and engineers use many functions in their work, such as square root, sine, and cosine, that are not basic operations on a computer and cannot be expressed in a finite number of basic operations. However, they can be approximated to any desired precision using basic operations, as we have seen in Chapters G3 and A1. Many of these functions can be described by means of *power series.* For example, we can write

$$\exp(x) = 1 + x + \frac{x^2}{2!} + \frac{x^3}{3!} + \ldots$$

$$\log(1 + x) = x - \frac{x^2}{2} + \frac{x^3}{3} - \frac{x^4}{4} + \ldots \qquad \text{(for } |x| < 1\text{)}$$

$$\sin(x) = x - \frac{x^3}{3!} + \frac{x^5}{5!} - \ldots$$

$$\cos(x) = 1 - \frac{x^2}{2!} + \frac{x^4}{4!} - \ldots$$

In principle, we can substitute the value of x into these power series and approximate the value of the function as closely as we like by computing a sufficient number of terms; but there are some difficulties, both theoretical and practical. Some series are valid only for restricted ranges of the argument x. The logarithm (log) function, for example, has no meaning if the absolute value of x is not less than one. (We can see this by considering the behavior of the series when, say, $x = 2$: the terms increase without bound. We say that the infinite sum *does not converge.*)

Even if the series is valid for all values of x, it may be possible to achieve greater precision with fewer terms when x is small. For example, if we evaluate exp(10.0) by substituting $x = 10.0$ into the series above, we find that we have to compute 25 terms to get four significant digits of precision. By comparison, if we evaluate exp(0.1), we can achieve four significant digits in just four terms. Consequently, we should never blindly evaluate the terms in the power series as they stand, but should look for ways to take advantage of the properties of the series. The standard technique is *argument reduction*, which attempts to reduce the argument to a small value. We will illustrate with several examples.

The series for the log function is valid only for a restricted range of argument values, so it is essential to reduce the argument to within this range. We can use the properties of logarithms to achieve this. Suppose the argument z is expressed as $z = f \times 10^e$, where $0.1 \leq f < 1.0$—that is, we have a decimal floating-point representation of the strictly positive number z. (The argument to the log function cannot be zero or negative: the logarithm does not then exist in real arithmetic.) By the properties of logarithms, we can write

$$\begin{aligned}\log(z) = \log(f \times 10^e) &= \log(f) + \log(10^e) \\ &= \log(1 + (f - 1)) + e \times \log(10)\end{aligned}$$

Since log(10) is a constant, we can evaluate $e \times \log(10)$ with a multiplication. Since $x = (f - 1)$ is between -0.9 and 0.0, we can approximate $\log(1 + (f - 1)) = \log(1 + x)$ by using the series above.

On computers using binary floating point, we can simply substitute 2 for 10 in the discussion above. In fact, we can do this even on a computer that uses some other form of number representation. To compute the logarithm of a number z, we first find an integer n such that

$$2^{n-1} \leq z < 2^n$$

Then we write

$$\log(z) = \log(2^{-n}z \times 2^n) = \log(1 + (2^{-n}z - 1)) + n \log(2)$$

Now the argument to be used in the series expansion of the log function is $x = 2^{-n}z - 1$, which is between -0.5 and 0.0. When an argument of -0.5 is used in the expansion, only 10 terms are needed to achieve four-digit precision, compared with nearly 60 terms when the argument is -0.9.

The argument of the exponential function can be reduced in a similar manner. Suppose that $z = n + f$, where n is an integer and f lies between -0.5 and $+0.5$. Then we can write

$$\exp(z) = \exp(n + f) = \exp(n) \times \exp(f)$$

The first term, $\exp(n)$, can be computed entirely with multiplications if n is positive, or with multiplications and one division if n is negative. The second term can be computed using the power-series approximation. Since the argument does not exceed 0.5 in magnitude, relatively few terms are needed. For example, four significant digits can be obtained with no more than six terms.

The arguments of the sine and cosine functions can be reduced by using their periodicity properties: that is, the fact that

$$\sin(x) = \sin(x + 2\pi n)$$
$$\cos(x) = \cos(x + 2\pi n)$$

for any integer n. (A *periodic function* with *period* T is a function f such that $f(x) = f(x + T)$ for all values of x. The period of the sine and cosine functions is 2π.) We can use this property to reduce any argument z to within the range 0 to 2π by writing

$$\sin(z) = \sin(z - 2\pi n)$$

where n is chosen so that $z - 2\pi n$ falls within the desired range. We can simplify the problem still further by using the relations

$$\sin(x - \pi) = -\sin(x)$$

$$\sin\left(x + \frac{\pi}{2}\right) = \cos(x)$$

The graph of sin(x) for x between 0 and 2π is shown in Figure A5.1. It is broken up into sections using the relationships above, so that we can use either the sine or the cosine series in the regions shown. For example, if x is between $5\pi/4$ and $7\pi/4$, we compute

$$\sin(x) = -\cos(x - 3\pi/2)$$

so that the argument of the cosine function satisfies

$$-\pi/4 \leq x - 3\pi/2 \leq \pi/4$$

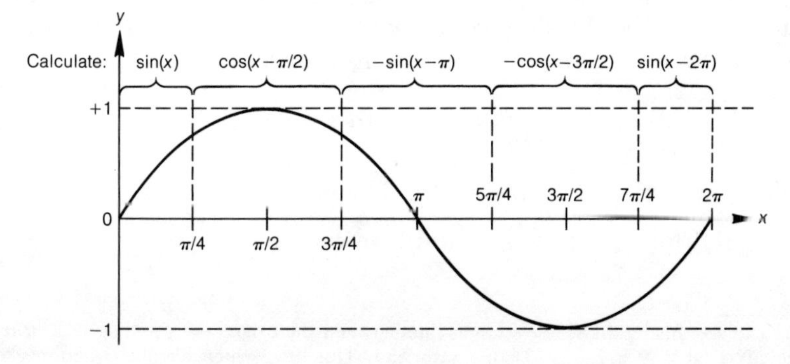

Figure A5.1 Calculation of sin(x)

In this way the argument used in the series expansion is always between $-\pi/4$ and $+\pi/4$.

Because we now know that $|x| \leq \pi/4 \cong 0.7854$, we can decide how many terms we need in the sine and cosine series. Suppose we want four decimal digits of precision: that is, we want

$$\left|\frac{\text{error in calculated } \sin(x)}{\sin(x)}\right| \leq 10^{-4}$$

It is known that $|\sin(x)| \geq \frac{2}{\pi} x$ for all $|x| \leq \frac{\pi}{2}$;* hence we ask that the sum of the neglected terms be less than $10^{-4}\,|x|\,\frac{2}{\pi}$. The series for sin(x) is an example of an *alternating series*: that is, the terms alternate in sign. Furthermore, when $|x| < 1$, the terms decrease in magnitude. If we were to approximate sin(x) by powers up to x^5, the truncation error would consist of the sum of those terms in parentheses in the series

$$\sin(x) = x - \frac{x^3}{3!} + \frac{x^5}{5!} - \left(\frac{x^7}{7!} - \frac{x^9}{9!}\right) - \left(\frac{x^{11}}{11!} - \frac{x^{13}}{13!}\right) - \ldots$$

The expressions in parentheses have the same sign as x, since $|x| < 1$. Hence the sum of the first three terms of the sine series gives a result whose error has the same sign as x: that is, the approximation is larger than sin(x) if x is positive and smaller than sin(x) if x is negative. We can also write

$$\sin(x) = x - \frac{x^3}{3!} + \frac{x^5}{5!} - \frac{x^7}{7!} + \left(\frac{x^9}{9!} - \frac{x^{11}}{11!}\right) + \left(\frac{x^{13}}{13!} - \frac{x^{15}}{15!}\right) + \ldots$$

This time the error in using the first three terms is seen to be $-x^7/7!$ plus a series of terms that have the same sign as x. Thus the error in using the first three terms is between 0 and $x^7/7!$. This is a particular instance of a general result for all alternating series of decreasing terms: *the error is no larger in magnitude than the first neglected term.* Consequently we can decide how many terms of the sine series to use by finding the smallest n for which the relation

$$\left|\frac{x^{2n+1}}{(2n+1)!}\right| < 10^{-4}\,|x|\,2/\pi$$

holds when $|x| \leq 0.7854$. This can be done by computing $(0.7854)^{2n}/(2n+1)!$ for $n = 1, 2, \ldots$ until we get a result less than $10^{-4}\,\frac{2}{\pi} \cong$

*We can see this by drawing a straight line between the origin and the first peak of the sine curve (at $x = \pi/2$, $y = 1$) in Figure A5.1. This line, which has the equation $y = \frac{2}{\pi}x$, is below the sine curve for all values of x between 0.0 and $\pi/2$.

TABLE A5.1 MAXIMUM ERROR FOR SINE

n	1	2	3	4
$\frac{(0.7854)^{2n}}{(2n+1)!}$	0.1028089	0.0031709	0.0000466	0.0000004

0.000064. The values are shown in Table A5.1, from which we see that using the first three terms of the sine series gives the desired precision.

To get four significant digits from the cosine series, we can again use the alternating property of the series. For $|x| \leq \pi/4$, $\cos(x) > 1/\sqrt{2} \cong 0.7071$. Hence we want the first neglected term to be less than 0.7071×10^{-4}. This time, therefore, we want to find an n such that $(0.7854)^{2n}/(2n)! < 0.7071 \times 10^{-4}$. Table A5.2 shows that an n of 4 is adequate. Therefore, for $|x| \leq \pi/4$, $\cos(x)$ can be approximated by

$$1 - \frac{x^2}{2!} + \frac{x^4}{4!} - \frac{x^6}{6!}$$

to four significant digits.

TABLE A5.2 MAXIMUM ERROR FOR COSINE

n	1	2	3	4
$\frac{(0.7854)^{2n}}{(2n)!}$	0.3084266	0.0158545	0.0003260	0.0000036

Program A5.1 illustrates the steps in the argument reduction. It first determines the number of multiples of 2π that can be subtracted from the input argument X. (In fact, it subtracts multiples of π. If an even number of multiples are subtracted, the result is not changed. If an odd number are subtracted, the sign of the result is reversed.) Program A5.1 calls on two auxiliary functions, SIN1 and COS1, which compute the sine and cosine using the power series, under the assumption that their arguments are no more than $\pi/4$. These functions could be written out in the obvious way, but there are some efficiency considerations, which will be taken up in Chapter A8.

Program A5.1 is *not* representative of the way the built-in functions SIN and COS are actually coded. Those functions are used very frequently, so it is important that they be made as efficient as possible. Normally they are written in machine language, to take advantage of every quirk of the computer, and use additional ideas to reduce the argument and hence the amount of arithmetic needed. However, Program A5.1 does illustrate the type of idea used.

Program A5.1 ***Argument reduction for sine computation***

```
SIN: subprogram (X)
    This function computes the sine of the argument X, which is given in
    radians. X is reduced until it is in the range 0.0 to π/4, and is then
    passed on to one of the auxiliary functions SIN1 or COS1.
    real SIN,X,X1,Z,PI=3.14159,PIBY2=1.57080,PIBY4=0.78540
    integer N
        N←X / PI
        X1←X−N*PI
        if X1>PIBY2 then X1←PI−X1 endif
        if X1<PIBY4
            then Z←SIN1(X1)
            else Z←COS1(PIBY2−X1)
            endif
        if N≠(N÷2)*2 then Z←−Z endif
        return (Z)
    endsubprogram SIN
```

Problems

1. Write a program to evaluate sin(x), using the power series without argument reduction and stopping when the next term is less than 10^{-6}. (See Program A1.2.) Run this program for $x = \pi/4, 9\pi/4, 17\pi/4, \ldots, (4n + 1)\pi/4$. Print the answer and the number of terms needed to achieve the required precision. Do you notice anything interesting in addition to the increased number of terms needed?
2. Write a program to compute exp(z), using argument reduction to get an argument in the range −0.5 to +0.5. Use an array E(I), containing exp(I) for I from 1 to 50, to assist in the computation. Assume that z does not exceed 50 in magnitude, and that four significant figures are needed in the answer.

Chapter

A6

Numerical Integration

A numerical problem that frequently has to be solved on a computer is that of *numerical integration*: finding the area or volume of an object whose shape is known. This problem can arise in relatively simple situations such as designing freeways, where it is necessary to know how much earth must be removed or filled in to bring the roadbed to the desired level, or in more complex problems such as airplane design, where the area of the wing affects (for example) the viscous drag. In this chapter we will briefly discuss the problem of computing numerical approximations to the areas of two-dimensional (that is, flat) objects. The extension of these results to computing the surface areas and volumes of three-dimensional (that is, solid) objects is not simple, but the basic ideas are similar.

Suppose we have a closed figure such as that shown in Figure A6.1, and we want to find its area. One method we often learn in grade school is to draw squares over the figure and count the number of squares, as shown in Figure A6.2. Suppose these squares are each one foot on a

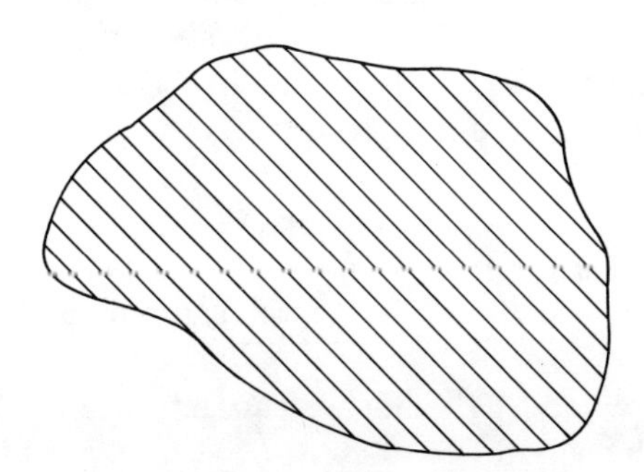

Figure A6.1 Closed figure

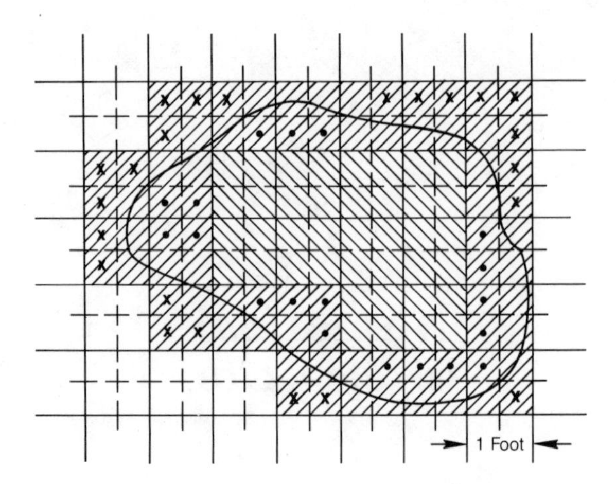

Figure A6.2 "Squares" method for finding area

side. The squares shaded [cross-hatched] are completely inside the figure. There are 10 of them, so the area is certainly no less than 10 square feet. The squares shaded [hatched] are partly inside and partly outside the figure: we will call these the *boundary squares.* There are 20 such squares, so the area of the figure is not more than 30 square feet (the sum of all the shaded squares). Thus we have the numerical result

$$10 \leq \text{area} \leq 30$$

Not a very precise estimate! How can we get more precision if we need it? One way is to use smaller squares. In Figure A6.2 we have also subdivided each square with dashed lines into 4 smaller squares. In this process the original 10 one-foot squares that were inside the figure have become 40 six-inch squares, all inside the figure. In addition we have another 19 six-inch squares (those with dots in their centers) that were previously part of the boundary squares and are now completely inside the figure. Thus the area of the figure is no less than $(10 \times 4 + 19)/4$ square feet. Of the remaining six-inch squares formed from the old one-foot boundary squares, 23 (those with x's in their centers) are now completely outside the figure. Thus the area is no greater than $(30 \times 4 - 23)/4$ square feet. We now have the numerical result

$$14.75 \leq \text{area} \leq 24.25$$

The uncertainty is still not very small, but it has improved by a factor of more than two (from 20 to 9.5). We can get more precision by making the squares even smaller.

We naturally wonder whether we can get as much precision as we like by making the squares as small as necessary. For this problem, the answer is *yes* (provided that the boundary line of the figure is reasonably smooth). In fact, for each factor of 2 (or any number α) that we decrease the side of the squares, the uncertainty will decrease by about the same factor. We can see this intuitively when we realize that the boundary squares will always include the boundary line of the original figure, and therefore form a "thick line" representation of the boundary that is about as thick as the squares. If the side of the squares is reduced, so is the thickness of the line. The area of the thick line is precisely the uncertainty, and is roughly proportional to the thickness of the line—that is, to the side of the squares.

The property that as we successively refine a process by taking more and more terms (in this example, squares) we get closer and closer to the true value is called *convergence.* Usually we try to use convergent processes, so that we can get as much precision as desired; but we must remember that the more terms we use, the more rounding errors will occur. Hence the sum of all the rounding errors will increase as the truncation error decreases. The total error is the sum of the two errors, as shown in Figure A6.3. We can see from the figure that the total error will initially decrease to a minimum, then increase again as the rounding error becomes dominant.

The measurement of an area by counting squares is about the crudest approximation we can use, but it does have the important property of convergence. Before investigating other methods, we will simplify the problem somewhat. In Figure A6.4 we have chopped the original figure up into five pieces with vertical and horizontal lines. The piece shaded diagonally is a rectangle, whose area can be calculated directly by multiplication. The remaining four pieces all have the general shape shown in Figure A6.5.

If we knew how to find the areas of regions like that in Figure A6.5, we could find the area of Figure A6.4 by adding up the areas of the five

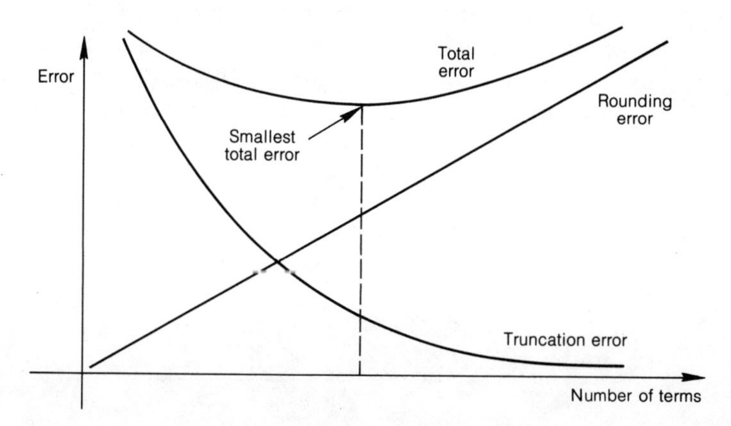

Figure A6.3 Total error

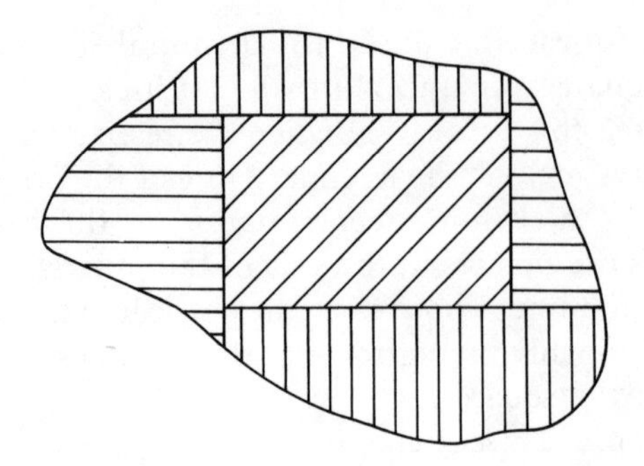

Figure A6.4 Division of region into subregions

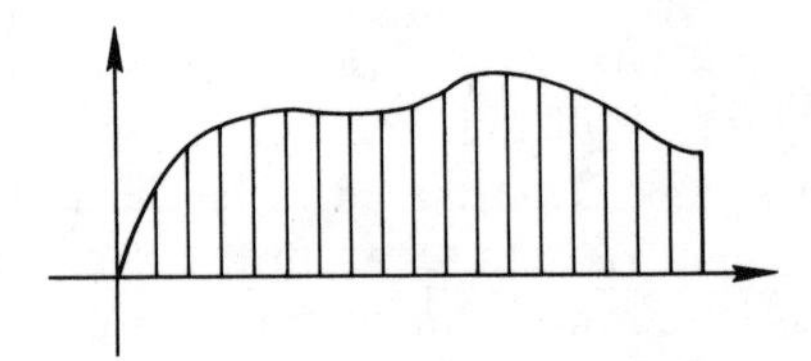

Figure A6.5 General shape of remaining subregions

subregions. Therefore we will study the restricted problem of finding the area below a curve $y = f(x)$ between the lines $x = A$ and $x = B$, as shown in Figure A6.6. This area is called the *definite integral of* $f(x)$ *between* A *and* B, and is written as

$$\int_A^B f(x)\,dx$$

If we put squares on Figure A6.6, we would be counting an integral number of squares to estimate the area, as shown in Figure A6.7a. Suppose, instead, that we use rectangles of the same width as the squares, but with heights determined by the function $f(x)$, as shown in Figure A6.7b. We can calculate the area of the rectangles easily, and we can see

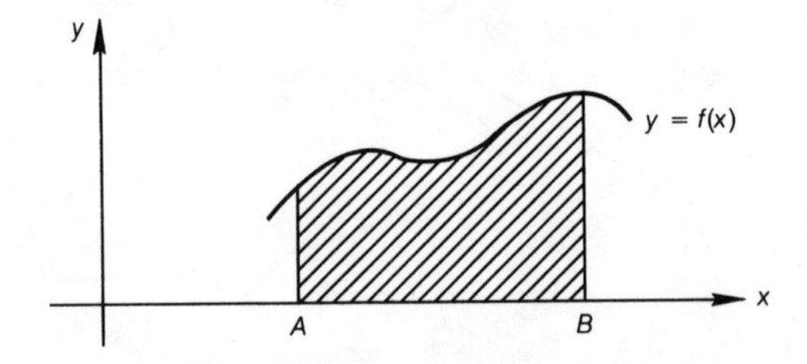

Figure A6.6 Definite integral of $f(x)$ between A and B

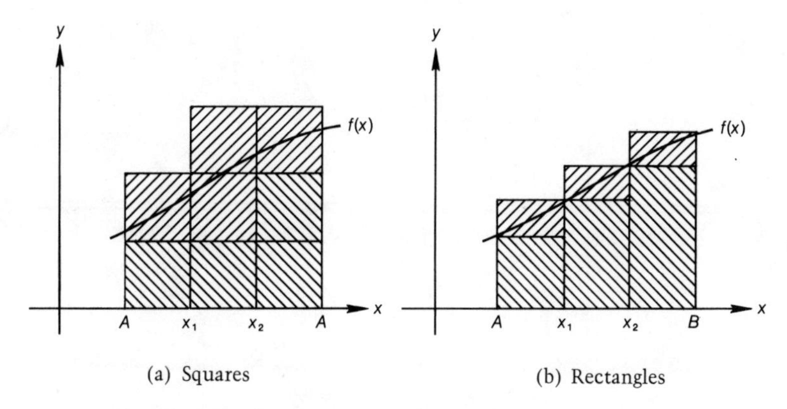

(a) Squares (b) Rectangles

Figure A6.7 Using squares and rectangles to determine area

that we will get at least as much precision. In Figure A6.7a the boundary region representing the uncertainty consists of four squares, whereas in Figure A6.7b it consists of three rectangles, each smaller than the corresponding square.

A6.1 TRAPEZOIDAL RULE

We have shown the interval $[A,B]$ chopped up into three smaller intervals in Figure A6.7. Let us write x_0 for A and x_3 for B. The rectangles all have the same width: let us call it h. Thus

$$x_1 - x_0 = x_2 - x_1 = x_3 - x_2 = h$$

or

$$x_n = x_0 + nh, \text{ for } n = 0, 1, 2, \ldots$$

The area shaded ▧ in Figure A6.7b consists of three rectangles. The height of the first is $f(x_0)$; the heights of the second and third are $f(x_1)$ and $f(x_2)$, respectively. Thus the total area, which is the sum of the areas of the three rectangles, is

$$\begin{aligned}(x_1 - x_0)\, f(x_0) + (x_2 - x_1)\, f(x_1) + (x_3 - x_2)\, f(x_2) \\ = h[f(x_0) + f(x_1) + f(x_2)]\end{aligned}$$

This is less than the area under the curve $f(x)$ in Figure A6.7.

Similarly, the total shaded area in Figure A6.7b, which is larger than the area under $f(x)$, is given by

$$\begin{aligned}(x_1 - x_0)\, f(x_1) + (x_2 - x_1)\, f(x_2) + (x_3 - x_2)\, f(x_3) \\ = h[f(x_1) + f(x_2) + f(x_3)]\end{aligned}$$

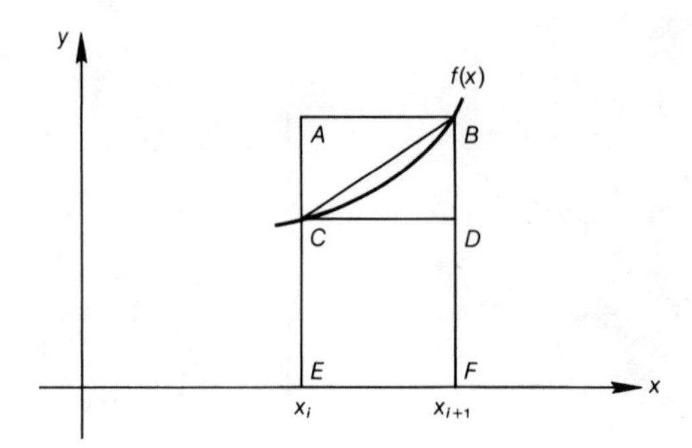

Figure A6.8 Trapezoidal rule

Since the first of these expressions is smaller than the actual area and the second is larger, we can take as a better approximation the average of the two. Thus we estimate the area as

$$\frac{1}{2}\Big[h[f(x_0) + f(x_1) + f(x_2)] + h[f(x_1) + f(x_2) + f(x_3)]\Big]$$

$$= \frac{h}{2}[f(x_0) + 2f(x_1) + 2f(x_2) + f(x_3)]$$

What are we really doing when we take the average of the two approximations? Figure A6.8 shows one of the rectangular regions. The smaller approximation uses the rectangular region *CDFE*, the larger uses *ABFE*. The larger, *ABFE*, is composed of *CDFE* and *ABDC*. Thus the average of the two areas is

$$\frac{1}{2}[\text{area}(ABFE) + \text{area}(CDFE)]$$

$$= \frac{1}{2}[(\text{area}(CDFE) + \text{area}(ABDC)) + \text{area}(CDFE)]$$

$$= \text{area}(CDFE) + \frac{1}{2}\text{area}(ABDC)$$

Half the area of *ABDC* is the area of the triangle *BDC*. Thus the average is the same as the area of the trapezoid *CBFE*, and our approximation is the area under *CB*, the *straight-line approximation* to the graph of $f(x)$ through *C* and *B*. This area is given by

$$\frac{(x_{i+1} - x_i)}{2}[f(x_i) + f(x_{i+1})] = \frac{h}{2}[f(x_i) + f(x_{i+1})]$$

This is called the *trapezoidal rule* for approximate integration.

We can take any interval, such as AB in Figure A6.6, chop it into subintervals, and apply the trapezoidal rule to each subinterval. For example, if we chop AB into three subintervals by dividing it at x_1 and x_2, we get

$$\text{area} \cong \frac{h}{2}[f(x_0) + f(x_1)] + \frac{h}{2}[f(x_1) + f(x_2)] + \frac{h}{2}[f(x_2) + f(x_3)]$$

$$= \frac{h}{2}[f(x_0) + 2f(x_1) + 2f(x_2) + f(x_3)]$$

as above. Similarly, if we chop the interval into N subintervals of length $h = (B - A)/N$, we get the approximation

$$\text{area} \cong \frac{h}{2}[f(x_0) + 2f(x_1) + 2f(x_2) + \ldots + 2f(x_{N-1}) + f(x_N)]$$

This method is also *convergent*: that is, as we make N larger, the approximation gets closer to the actual area.

Example A6.1 *Area of a Circle by the Trapezoidal Rule*

We wish to compute the area of a circle of radius 1. It can be divided into four equal segments and a square, as shown in Figure A6.9. Since the diagonal of the square is twice the radius of the circle, the side of the square is $\sqrt{2}$. Its area is therefore $(\sqrt{2})^2 = 2$. The areas of the four segments must be approximated by numerical integration. One segment is shown in Figure A6.10. The equation of the curved boundary is $(y + \sqrt{2}/2)^2 + x^2 = 1$, because the center is at $x = 0$, $y = -\sqrt{2}/2$ (the point O in the figure).

Because AB, the side of the square, is $\sqrt{2}$, the point A is ($x = -\sqrt{2}/2$, $y = 0$), and the point B is ($x = +\sqrt{2}/2$, $y = 0$). V is the point ($x = 0$, $y = 1 - \sqrt{2}/2$). The midpoints of AQ and QB are also shown. P is ($x = -\sqrt{2}/4$, $y = 0$) and R is ($x = \sqrt{2}/4$, $y = 0$). Hence U and W have a y-coordinate of $\sqrt{7/8} - \sqrt{2}/2$.

We will apply the trapezoidal rule, first to the two intervals AQ and QB, then to the four intervals AP, PQ, QR, and RB. The approximate

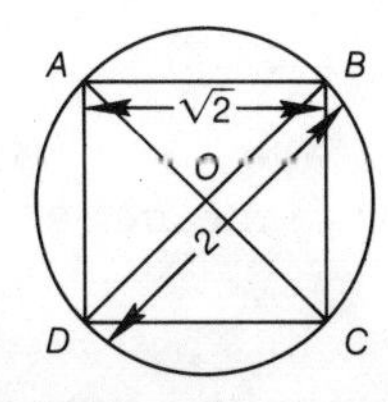

Figure A6.9 Five subregions of a circle

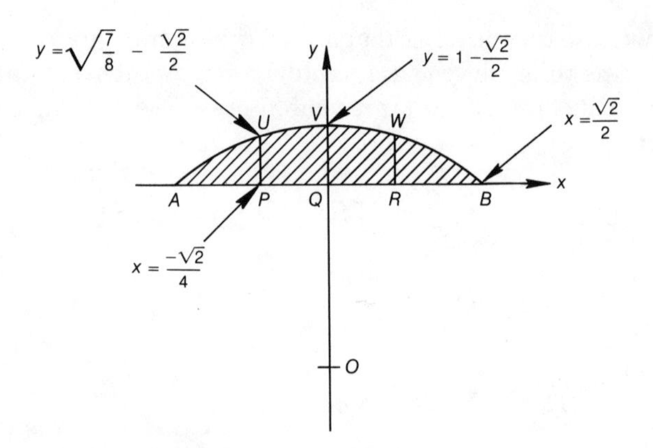

Figure A6.10 One segment of the circle $y = \sqrt{1 - x^2} - \sqrt{2}/2$

area of the circle will then be four times our approximation to the area of the segment plus the area of the square, which we know to be 2.

Approximating with two intervals,

$$\text{area} \cong \frac{AQ}{2}[y(A) + 2y(V) + y(B)]$$

$$= \frac{\sqrt{2}}{4}(0 + 2 - \sqrt{2} + 0) = 0.207$$

This leads to an approximation of $2 + 4 \times 0.207 = 2.828$ for the area of the circle.

Repeating the calculation with four intervals,

$$\text{area} \cong \frac{AP}{2}[y(A) + 2y(U) + 2y(V) + 2y(W) + y(B)]$$

$$= \frac{\sqrt{2}}{8}[0 + 2(\sqrt{7/8} - \sqrt{2}/2) + 2(1 - \sqrt{2}/2) + 2(\sqrt{7/8} - \sqrt{2}/2) + 0]$$

$$= \frac{\sqrt{2}}{8}(4\sqrt{7/8} - 2\sqrt{2} + 2 - \sqrt{2})$$

$$= 0.265$$

This gives an approximation to the area of the circle of $2 + 4 \times 0.265 = 3.060$.

The area of a circle of radius 1 is $\pi \cong 3.142$. Thus the errors are $3.142 - 2.828 = 0.314$ for the two-interval case, and $3.142 - 3.060 = 0.082$ for the four-interval case. When we used the square-counting method, we saw that halving the size of the squares just about halved the error. This time we see that, with the trapezoidal rule, the error is divided by about four when the interval is halved. If we halve the interval again, the error is about 0.022.

A6.2 SIMPSON'S RULE

Simpson's rule is another, more accurate method for approximating areas. We will derive it intuitively from the trapezoidal rule, without going into any theory. Simpson's rule is applicable to many problems, and is the basis of many numerical-integration programs available in computer libraries. If you ever need to perform numerical integration, consider using one of these library programs. You should not attempt to write your own program for problems of this type unless there are very good reasons why the library versions are not applicable to your program: carefully developed library programs are much more likely to be error-free and to look after some of the messy details that we cannot discuss in an introductory presentation.

We commented at the end of the previous section that the error in the trapezoidal rule reduces by approximately one-fourth when the number of intervals is doubled. Consider two numerical approximations to the integral of $f(x)$ over the interval AB, first using one interval and then using two half-length intervals, as shown in Figure A6.11. Let us

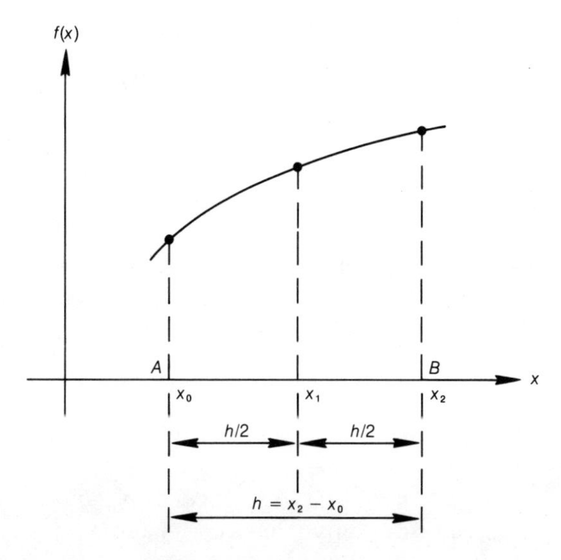

Figure A6.11 Simpson's rule

call these approximations A_1 and A_2. They are

$$A_1 = \text{trapezoidal rule using one interval of length } h = x_2 - x_0$$
$$= \frac{h}{2}[f(x_0) + f(x_2)]$$

$$A_2 = \text{trapezoidal rule using two intervals of length } h/2$$
$$= \frac{h}{4}[f(x_0) + 2f(x_1) + f(x_2)]$$

Suppose that the error in A_2 is e: that is,

$$A_2 = \text{area} + e$$

Since the error in A_1 is about four times the error in A_2, we have the approximation

$$A_1 = \text{area} + 4e$$

If we ignore the fact that the last equation is only an approximation, we can solve the two equations

$$A_1 = \text{area} + 4e$$
$$A_2 = \text{area} + e$$

for the unknown area and hope to obtain a better approximation than we can get by the trapezoidal rule. By subtracting 4 times the second equation from the first and dividing by -3, we obtain

$$\text{area} = (4A_2 - A_1)/3$$

Replacing the values of A_1 and A_2 with the expressions given above, we get

$$\text{area} = \frac{h}{6}[f(x_0) + 4f(x_1) + f(x_2)]$$

This is called *Simpson's rule* for approximate integration.

As with the trapezoidal rule, we can break an interval up into many subintervals and apply Simpson's rule to each one. Thus, if we choose $h = (B - A)/N$ and $x_{i+1} - x_i = h/2$, with $x_0 = A$ and $x_{2N} = B$, we have

$$\int_A^B f(x)\,dx = \text{area under curve } f(x) \text{ between } A \text{ and } B$$

$$= \text{area under curve } f(x) \text{ between } x_0 \text{ and } x_2$$
$$+ \text{ area under curve between } x_2 \text{ and } x_4$$
$$+ \ldots$$
$$+ \text{ area under curve between } x_{2N-2} \text{ and } x_{2N}$$

$$\cong \frac{h}{6}[f(x_0) + 4f(x_1) + f(x_2)]$$

$$+ \frac{h}{6}[f(x_2) + 4f(x_3) + f(x_4)]$$

$$+ \ldots$$

$$+ \frac{h}{6}[f(x_{2N-2}) + 4f(x_{2N-1}) + f(x_{2N})]$$

$$= \frac{h}{6}[f(x_0) + 4f(x_1) + 2f(x_2) + 4f(x_3) + 2f(x_4)$$

$$+ \ldots + 2f(x_{2N-2}) + 4f(x_{2N-1}) + f(x_{2N})]$$

Example A6.2 ***Area of a Circle by Simpson's Rule***

We will use Simpson's rule to calculate the area of a unit circle. As before, we break the circle into a square of area 2 and four equal segments like the one in Figure A6.10.

First we use the single interval AB shown in Figure A6.10.

$$\text{area} \cong \frac{\sqrt{2}}{6}[y(A) + 4y(V) + y(B)]$$

$$= \frac{\sqrt{2}}{6}[0 + 4 - 2\sqrt{2} + 0]$$

$$= 0.276$$

This gives the numerical approximation to the area of the circle as $2 + 4 \times 0.276 = 3.104$.

Next we use the two intervals AQ and QB to get

$$\text{area} \cong \frac{\sqrt{2}}{12}[y(A) + 4y(U) + 2y(V) + 4y(W) + y(B)]$$

$$= \frac{\sqrt{2}}{12}[0 + 4\sqrt{7/8} - 2\sqrt{2} + 2 - \sqrt{2} + 4\sqrt{7/8} - 2\sqrt{2} + 0]$$

$$\cong 0.2843$$

This gives a numerical approximation to the area of the circle of $2 + 4 \times 0.2843 = 3.1372$. The errors in the two cases are about 0.038 and 0.0044. This time the error has been divided by about 9. If we divided the interval into two again, we would find that the error reduces by about 16 each time the interval is halved, once the interval is small enough.

Program A6.1 is a function subprogram that computes an approximation to the integral of the function f, using Simpson's rule on the interval (A,B) divided into N equal subintervals. The parameter F is the name of a function that computes the value f.

Program A6.1 ***Integration by Simpson's rule***

```
INTEGRATION: subprogram (A,B,F,N)
   The real function F is integrated over the interval (A,B) using
   Simpson's rule. The interval is divided into N equal subintervals.
   real A,B,INTEGRATION,F,H,H2,SUM1,SUM2,X,Z
   integer I,N
      if N≤0
         then
            output 'INTEGRATION CALLED WITH N≤0'
            Z=0.0
         else
            H←(B−A) / FLOAT(N)
            H2←H / 2.0
            SUM1←0.0
            SUM2←F(A+H2)
            do for I←1 to N−1
               X←A+I*H
               SUM1←SUM1+F(X)
               SUM2←SUM2+F(X+H2)
               enddo
            Z←H*(F(A)+2.0*SUM1+F(B)+4.0*SUM2) / 6.0
         endif
      return (Z)
   endsubprogram INTEGRATION
```

Problem

1. Write a function subprogram that takes the same parameters as Program A6.1, but uses the trapezoidal rule to compute the approximation.

Chapter

A7

The Monte Carlo Method

The name *Monte Carlo method* comes from the famous gambling resort. In any game of chance, some device—such as a roulette wheel or a shuffled deck of cards—generates one of a set of random outcomes, and the players bet on that outcome. In roulette, for example, the standard wheel is marked with 38 different numbers, each of which is equally likely to occur when the wheel is spun. Consequently, the *probability* that any given number will come up is 1 in 38. If a player bets on a particular number, the bet will be lost 37 times and won once (on the average). Since the house pays less than 37-to-1 odds for a win, the player is bound to lose if the game is continued long enough, and the house is bound to win. (The house, unlike the player, is not there for the fun of it.) No betting strategy, or "system," can overcome the odds forever—eventually the "law of averages" will catch up with the player.

In statistics, the "law of averages" is called the *law of large numbers*, and states that as the number of bets increases, the likelihood that the outcome will deviate significantly from the *expected outcome* gets smaller and smaller. Thus if the payoff on the roulette wheel is 35 to 1—that is, out of 38 bets the player expects to lose a dollar 37 times and win 35 dollars once—the expected loss to the player is 2/38 of a dollar per play. This is the house profit. Betting on the "reds" does not change this. (Eighteen of the numbers are colored red, and a player can bet on red at even money. That is, if any of the red numbers comes up, a dollar is won, otherwise a dollar is lost.) Since, on the average, 18 of 38 spins will come up red, 18 dollars are won and 20 are lost, for an expected loss of 2/38 of a dollar per play. No matter what combination of bets is made or what the amount of the bets, fixed or varied, in the long run the player will lose 2/38 of the total amount bet (in the sense that the difference between the amount actually lost and this expected loss will get progressively smaller compared with the total amount bet).

This property of random numbers means that they can be used to solve certain types of problem. Section A11.2 discusses discrete simulation, which uses random numbers to simulate the behavior of random processes. We use this technique to study the expected behavior of a system over a long period of time. For example, we might want to determine the average flow of traffic through a city street as a function of the timing of the traffic lights. If we had a game of chance that was more difficult to analyze than roulette, we might consider estimating the probabilities of various outcomes by taking statistics. (Indeed, this may not be a bad idea, even in roulette: it is one way of determining whether the wheel is properly adjusted or has been set to favor the house!)

The *Monte Carlo method* is a method that uses random numbers to solve nonrandom problems. Suppose, for example, that we have a square board with a circle inscribed in it, as shown in Figure A7.1. If we throw darts at this board so that they can land anywhere on the board with equal likelihood, what is the probability that a dart will land inside the circle? Clearly, it is the ratio of the area of the circle to the area of the square: or $\pi/4$, or about 0.7854. If we did not know the value of π, we could estimate it by throwing a large number of darts at the board and computing the proportion that land inside the circle. This technique can be implemented on a computer as a way of computing the area of a complex region.

In Chapter A6 we developed several *deterministic* methods for computing areas, and applied two of them to the problem of computing the area of a circle. Normally we do not use Monte Carlo, or *probabilistic*, methods when deterministic methods are practical. However, there are some integration problems that are very difficult and time-consuming to handle deterministically, and for these we must often resort to Monte Carlo methods. We will again illustrate by computing the area of a circle.

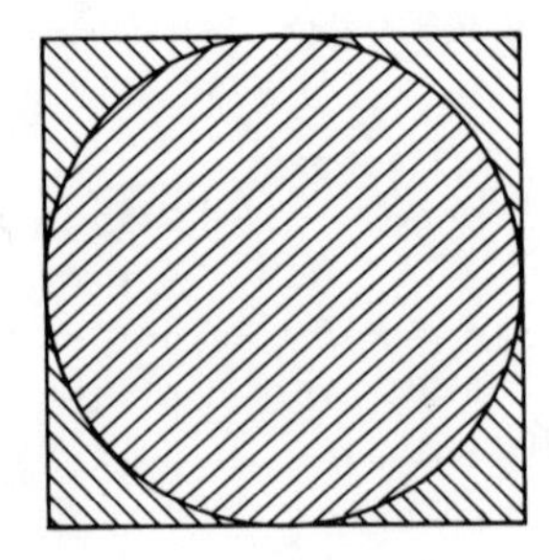

Figure A7.1 Circle inscribed in a square

Most computers provide subprograms for generating sequences of numbers that appear to be random. We say "appear to be" because such a sequence is generated by following a prescribed rule that determines the next number from those that have gone before. (We say that the sequence is *deterministic.*) A typical rule used to generate random integers is to multiply the previous integer by a carefully chosen number, and then to extract some of the digits from the result as the next number in the sequence. However, if we ignore the fact that there is an underlying generating procedure, the numbers generated by such a program have all the properties of random numbers. For the purpose of further discussion, we will assume that we have a function subprogram, RANDOM(), which returns a real value betweeen 0.0 and 1.0, all values being equally likely. (RANDOM is a function of no arguments, but we follow it with an empty pair of parentheses to make it clear that it is a function and not a simple variable.)

A computer program to estimate the value of π by "throwing darts"

Program A7.1 ***Compute π by Monte Carlo method***

```
MONTE__CARLO: program
    The value of π is estimated by taking random samples (X, Y) in a
    unit square and counting the number of samples that lie inside a
    unit circle centered at the origin. The probability that a sample is
    in the unit circle is π/4.
    integer I,M,N,COUNT
    real ESTIMATE,X,Y
        N←0
        COUNT←0
        output 'INPUT NUMBER OF THROWS TO BE MADE'
        input M
        do while M>0
        |   do for I←1 to M
        |   |   X←RANDOM()
        |   |   Y←RANDOM()
        |   |   if X*X+Y*Y<1.0 then COUNT←COUNT+1 endif
        |   |   enddo
        |   N←N+M
        |   ESTIMATE←FLOAT(COUNT) / FLOAT(N)*4.0
        |   output N,ESTIMATE,ESTIMATE−3.1416
        |   output 'INPUT NUMBER OF ADDITIONAL THROWS TO
        |      BE MADE'
        |   input M
        |   enddo
    endprogram MONTE__CARLO
```

TABLE A7.1 SAMPLE OUTPUT FROM PROGRAM A7.1

N	ESTIMATE	ERROR
100	3.2000	0.0584
200	3.1000	−0.0416
300	3.0267	−0.1149
400	3.0700	−0.0716
500	3.0560	−0.0856
1000	3.0760	−0.0656
2000	3.1380	−0.0036
3000	3.1707	0.0291
4000	3.1570	0.0154
5000	3.1720	0.0304
10000	3.1596	0.0180
20000	3.1580	0.0164
30000	3.1521	0.0105
40000	3.1464	0.0048
50000	3.1536	0.0120
100000	3.1443	0.0027

at Figure A7.1 is given in Program A7.1. The program expects the user to input an integer representing the number of "darts" to be "thrown." Each throw is simulated by obtaining two random numbers between zero and one, representing the X- and Y-coordinates of the point at which the dart lands. (Since X and Y are both between 0 and 1, the dart always lands in the *positive quadrant*—that is, the upper right-hand corner—but since the full figure can be constructed from four identical pieces, the result is not changed The dart will still land inside the circle with probability $\pi/4$.) As each position is computed, a counter is increased by one if the position falls inside the circle (that is, if $X^2 + Y^2 < 1.0$). When the number of "throws" requested has been made, the program prints the estimate of π obtained by taking four times the proportion that fall inside the circle. The process may then be repeated for an additional number of throws. Table A7.1 shows the result of a sample execution of this program. Notice that the result approaches the value of π very slowly. The Monte Carlo method is not a good method to use if direct methods are available; rather, it is used in very complex problems that are difficult to analyze directly and therefore difficult to solve by any other method.

Problems

1. The earth is an *oblate spheroid*—that is, a sphere flattened at the poles. In this problem we will grossly exaggerate the flattening by assuming that the interior of the earth is the set of points (X, Y, Z)

such that $X^2 + Y^2 + 1.2\,Z^2 < 1.0$, where X, Y, and Z are expressed in units of the earth's equatorial radius (the distance from the center of the earth to the equator, 6.37816×10^6 meters). If points are chosen randomly in the unit cube $0.0 \leq X, Y, Z \leq 1.0$, then the probability that a point is inside the earth is equal to the volume of the earth divided by the volume of the unit cube. Write a program to estimate the volume of the flattened earth using the Monte Carlo method.

If you want an accurate figure for the volume of the earth, change the coefficient from 1.2 to 1.00674 in the inequality above. The answer must be multiplied by the radius cubed, 2.59469×10^{20}, to get the volume in cubic meters. Unfortunately, an enormous number of samples must be used to get any appreciable accuracy—it is easier to look the answer up! According to the Encylcopedia Britannica, the volume is $1.083218915 \times 10^{21}$ cubic meters.

2. To make a random selection among a small number of possibilities, a random-number generator can be used as follows: Suppose we want to throw a six-sided die to get a result that is equally likely to be any integer between 1 and 6. The value of `1.0+6.0*RANDOM()` will be a random real number between 1.0 and 7.0. Truncated to an integer, the result will be one of the integers 1 through 7—though 7 should hardly ever occur. (Many random-number generators generate results *greater than or equal to* zero but strictly *less than* one, so 7 will *never* occur.) If 7 occurs, it should be discarded and another random number generated. Otherwise, the result will be the desired random integer. Using this idea, write a program to estimate the probability that a "full house" will be thrown with five dice. (A full house consists of three dice showing the same value, and the other two showing an equal but different value: for example, 3, 5, 3, 3, 5.) At the same time, compute the probability of "three of a kind" (three dice show the same value, but the other two are different).

Chapter

A8

Evaluation of Functions

Engineering, scientific, and economic calculations often require the evaluation of various functions. These may be well-known mathematical functions, such as sine and exponential, or they may be functions peculiar to the problem, such as the force needed to stretch a particular type of material a given amount. Most functions that are needed cannot be defined exactly in terms of the basic operations of addition, subtraction, multiplication, and division. However, as we have seen in Chapter A5, many functions can be defined by means of power series, and approximated by calculating the first few terms of those series. To calculate functions in this way, we need to be able to evaluate polynomials such as $p(x) = a_0 + a_1x + a_2x^2 + \ . \ . \ . \ + a_nx^n$ efficiently. *Horner's method,* described in section A8.1, is the fastest way of evaluating such polynomials.

In other cases we have no knowledge of the underlying mathematical form of the function, but we have experimentally determined its value at a number of points. For example, suppose we have measured the stretching (strain) of a metal bar for various amounts of applied force (stress), and have obtained the values shown in Figure A8.1. We might want to know the strain for values of the stress that we haven't tried. If we assume that a smooth curve can be drawn connecting the measured points, as in Figure A8.1, we can obtain the desired values from that curve. If this information is to be used in a computer program (for example, to calculate whether a proposed design for a bridge will stand up to intended usage), we would like to store the observed values in the computer, and use them to calculate an approximation to the strain for any applied stress. Section A8.2 will discuss one scheme for doing this, *linear interpolation,* which uses the approximation formed by

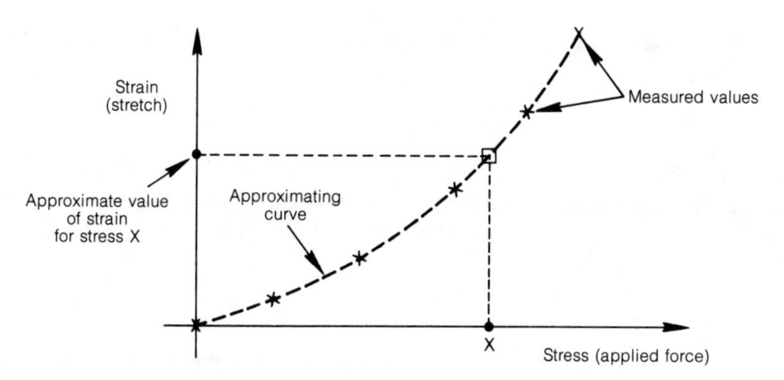

Figure A8.1 Measured values of a function

drawing straight lines between adjacent observed points. More accurate approximations can be obtained by using different types of curve, such as *cubic* curves (of the form $a_0 + a_1x + a_2x^2 + a_3x^3$), but the theory of such approximations is not an appropriate topic for discussion here.

A8.1 HORNER'S METHOD

The obvious way of writing an expression for a polynomial of *degree* n implies the use of $n(n + 1)/2$ multiplications. For example, the cubic (degree-3) polynomial $a_0 + a_1x + a_2x^2 + a_3x^3$ implies six multiplications. This can be seen if the multiplications are indicated explicitly:

$$p(x) = a_0 + a_1*x + a_2*x*x + a_3*x*x*x$$

We can save one multiplication by forming x^3 as the product of x and x^2, which we have already calculated. In this way we can calculate a polynomial of degree n with only $2n - 1$ multiplications, by writing it in the form

$$p(x) = a_0 + a_1*x + a_2*x_2 + a_3*x_3 + \ . \ . \ . \ + a_n*x_n \quad (n \text{ multiplications})$$

where

$$\left.\begin{array}{l} x_2 = x*x \\ x_3 = x*x_2 \\ . \\ . \\ . \\ x_n = x*x_{n-1} \end{array}\right\} \quad (n - 1 \text{ multiplications})$$

This calculation could be programmed using a loop. However, the number of multiplications can be reduced still further—to a total of

n—by factoring the polynomial. The cubic example given above can be written as

$$p(x) = a_0 + x*(a_1 + x*(a_2 + x*a_3))$$

which takes only three multiplications (and no more additions than the previous methods). This form of factorization is known as Horner's method, and it can be shown that it is not possible, in the general case, to reduce the number of multiplications or additions further.

If a low-degree polynomial is to be evaluated, a single expression can be written using this factored form; but if the polynomial is of high degree, it is simpler to use a loop. Program A8.1 gives a function subprogram to evaluate a polynomial of degree N whose Ith coefficient is in location A(I+1). Notice that the loop starts with the Nth coefficient, in location A(N+1), and works backward to the lowest-degree term.

Program A8.1 ***Evaluate polynomial by Horner's method***

```
POLYNOMIAL: subprogram (A,N,X)
    Function to compute the value of the polynomial

        A(1)+A(2)*X+ . . . +A(N+1)*X↑N

    using only N multiplications.
    integer I,N
    real POLYNOMIAL,A(N+1),P,X
        P←A(N+1)
        do for I←N to 1 by -1
            P←P*X+A(I)
            enddo
        return (P)
    endsubprogram POLYNOMIAL
```

If some of the powers of x are missing from the polynomial, it may be possible to improve the speed of the program. The power-series approximation to the cosine function, for example, includes only even powers of x. We can approximate the cosine by using the first seven terms in the form of an expression:

```
1.0+Y*(-0.5+Y*(0.41667E-1+Y*(-0.13889E-2
    +Y*(0.24802E-4+Y*(-0.27557E-6+Y*0.20877E-8)))))
```

where Y=X*X. (The coefficients are $-1/2!$, $1/4!$, $-1/6!$, $1/8!$, $-1/10!$, and $1/12!$.)

Program A8.2 performs the same computation using a loop instead of a single expression. If the function COS(X) were not provided in a language, we could define it by writing a function subprogram like the one shown. Notice that this program uses an initial-value declaration (discussed fully in Section P9.5) to initialize the array C. The seven values given in parentheses in the declaration statement, following the equal sign, will be placed into the seven locations in array C when the program is first loaded into memory. Since the elements of C are not changed during execution, they will have these values whenever the subprogram is used.

Program A8.2 ***Compute cosine by power series***

```
COS: subprogram (X)
    Array C contains the coefficients for the first seven terms in the
    power series: 1/0!, −1/2!, 1/4!, −1/6!, 1/8!, −1/10!, and 1/12!.
    real C(7)=(1.0,−0.5,0.41667E−1,−0.13889E−2,
      0.24802E−4,−0.27557E−6,0.20877E−8)
    real X,Y,P,COS
    integer I
        P←C(7)
        Y←X*X
        do for I←6 to 1 by −1
            P←P*Y+C(I)
            enddo
        return (P)
    endsubprogram COS
```

A8.2 LINEAR INTERPOLATION

If a function is defined by means of *tabular data*—that is, if we are given a table containing the values of the function Y(I) for a set of points X(I)—we must use some form of *interpolation* to approximate the value of the function for values of X that are not tabulated. *Linear interpolation* consists in approximating the function between any pair of points by a straight line, as shown between the points P and Q in Figure A8.2. If we connect each pair of consecutive points in our table by this method, we get an approximation to the curve of the function such as that shown in Figure A8.3.

To approximate the value of the function for an X value not in the table, we must make a calculation based on the geometry shown in Figure A8.2. First we must find the pair of consecutive X values X(I) and X(I+1) that surround the desired point X. Then we can compute the value Y of the corresponding point on the straight line PQ as

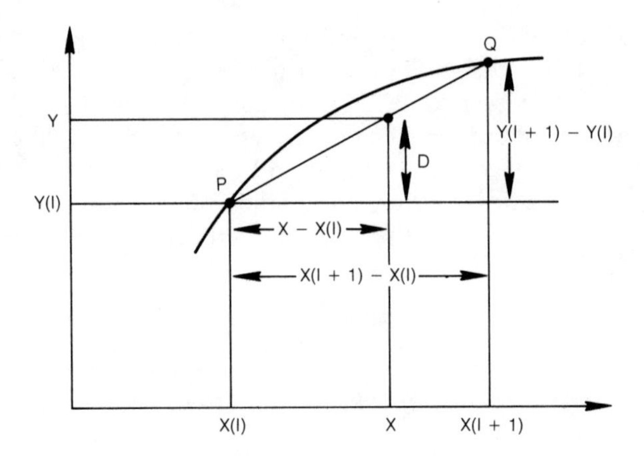

Figure A8.2 Computing the linear approximant

$$Y = Y(I) + D$$

$$= Y(I) + [Y(I+1) - Y(I)]\frac{X - X(I)}{X(I+1) - X(I)}$$

There are two different cases to consider: equal and unequal spacing. If the points X are equally spaced—that is, if

$$X(2) - X(1) = X(3) - X(2) = \ . \ . \ . \ = X(N) - X(N-1)$$

then we do not have to store the values of X(I): if we call the common spacing

$$H = X(2) - X(1) = X(I+1) - X(I)$$

and define

$$X0 = X(1) - H$$

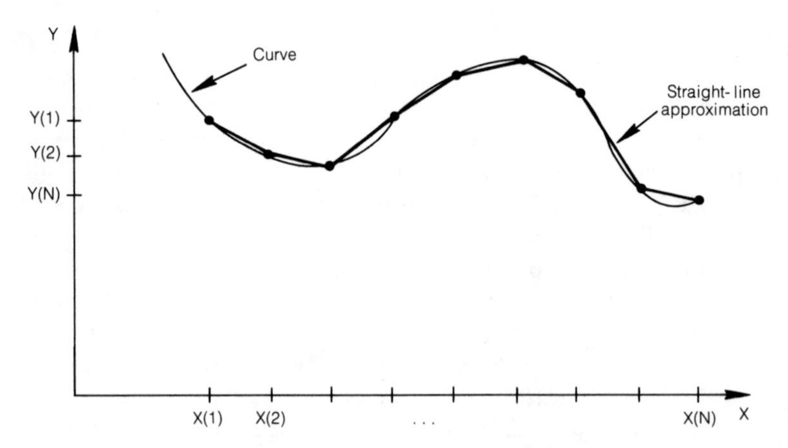

Figure A8.3 Linear interpolation

then we can compute X(I) from the expression

$$X(I) = X0 + I \times H$$

The appropriate I for any given value of X can be found by subtracting X0, dividing the remainder by the size of the intervals H, and truncating the result to an integer. That is, we can perform the following steps, using a real variable R and an integer variable I:

```
R←(X−X0) / H
I←R
```

Since we have defined $H = X(I+1) - X(I)$, the formula given above for approximating Y by linear interpolation reduces to

$$Y = Y(I) + [Y(I+1) - Y(I)]\frac{X - X(I)}{H}$$

The first term, Y(I), can simply be read out of our table; the factor $\frac{X - X(I)}{H}$ needed to calculate the second term can be found easily by noting that

$$\frac{X - X(I)}{H} = \frac{X - X0}{H} - \frac{X(I) - X0}{H} = R - I$$

A complete function subprogram for calculating Y by this method is shown in Program A8.3.

Program A8.3 ***Linear interpolation***

LINEAR__INTERP: **subprogram** (Y,X0,H,N,X)

Function subprogram to approximate a function by linear interpolation. The function value in array element Y(I) corresponds to the argument value

$$X(I) = X0 + I*H$$

If the value supplied for X is out of range, a warning message is printed and the value of the nearest endpoint is returned as the value of the function.

```
integer I,N
real Y(N),X0,H,X,Z,R,LINEAR__INTERP
    R←(X−X0) / H
    I←R
    if I<1 or I≥N
        then
            output 'X VALUE OUT OF RANGE'
```

```
            if I<1
                then Z←Y(1)
                else Z←Y(N)
                endif
        else
            R←R−I
            Z←Y(I)+(Y(I+1)−Y(I))*R
        endif
    return (Z)
endsubprogram LINEAR__INTERP
```

Higher-order interpolation is usually used to achieve greater accuracy or greater smoothness. If we are interested only in the closeness of the approximation, we usually use more than two points to determine a nonlinear polynomial function that passes through those points. For example, we can take three adjacent points in Figure A8.3, say the points on the curve corresponding to X=X(I−1), X=X(I), and X=X(I+1), and find the coefficients of a quadratic polynomial $AX^2 + BX + C$ that passes through those points. Then, in the neighborhood of X=X(I), we can approximate the function by that quadratic polynomial. If the accuracy is still insufficient, we can determine a cubic (third-order) polynomial by using four points.

At times we are also concerned with the smoothness of the approximation. Ship and automobile designers, for example, are interested in approximating smooth curves for the surfaces of hulls and fenders. In this case, higher-order polynomials are used to avoid the sharp bends at the joints of the *piecewise-linear* approximation in Figure A8.3. The curves given by these polynomials are similar to the "French curve" used in drafting to draw a smooth line connecting a number of points. Such curves, called *splines*, are very important in computer graphics and computer-aided design, but, unfortunately, are beyond the scope of this text.

If the points are not equally spaced, an array of X values must be supplied along with those for Y. The program must then search for a value of I such that $X(I) \leq X < X(I+1)$. Since the X values are stored in order, this can be done with a fast binary search (see Section A3.1) if there are enough points to warrant it. In some applications, a particular function will be calculated for many neighboring values of X. In that case, we can use the previous value of I as a first guess for the value of I to be used in the next approximation. Such a program would operate by first checking to see whether the value of X is between X(I) and X(I+1) for the last I used. If not, a sequential search forward or backward can be used to find the appropriate I.

Problems

1. Program a linear-interpolation function for unequally spaced points. The parameters should be arrays X and Y, each of dimension N, an X value XV (we are already using the name X for one of the arrays), and the integer N.

*2. Suppose you are given two N-element arrays, X and Y, containing values of a function Y corresponding to argument values X. Suppose further that the signs of Y(1) and Y(N) differ, and that the values X(I) are in increasing order as I increases. Write a program to find a value XO such that the corresponding value of the function is zero, assuming that the function can be approximated by linear interpolation between the tabulated values.

3. Suppose you are given a function tabulated at equal intervals: that is, the array element Y(I) gives the value of the function for X= XO+H*I, I from 1 to N. Write a subprogram to find all zeroes of the function, if any, between XO+H and XO+N*H by linear interpolation. (A *zero* of the function is a value of X for which the corresponding value of the function is zero.) Your subprogram should leave the number of zeroes found in an output parameter M, and the values of the solution in an array ZERO(I), for I from 1 to M. (You will have to dimension ZERO by N − 1, as there may be that many solutions.)

Chapter

A9

Solution of Linear Equations

We all know how to solve a pair of simultaneous equations in two unknowns, such as

$$\begin{aligned} 3x + 6y &= 15 \\ 2x + 3y &= 8 \end{aligned}$$

We divide the first equation by 3 (the coefficient of x) to get

$$x + 2y = 5$$

which expresses the value of x in terms of the value of y. Then we substitute this value of x into the second equation to get an equation for y only. One way of doing the second step is to multiply the equation obtained from the first step by the coefficient of x in the second equation (in this case 2) to get

$$2x + 4y = 10$$

and subtract this from the second equation, getting

$$-y = -2$$

or $y = 2$. We can now substitute this value back into the equation for x to find that $x = 1$. These steps are tabulated below.

Initial equations	$\begin{cases} 3x + 6y = 15 \\ 2x + 3y = 8 \end{cases}$	(a) (b)
Divide eq. (a) by 3	$x + 2y = 5$	(c)
Eq. (c) now replaces eq. (a)		
Multiply eq. (c) by 2	$2x + 4y = 10$	(d)
Subtract eq. (d) from eq. (b)	$-y = -2$	(e)
Divide eq. (e) by -1	$y = 2$	(f)

Eq. (f) replaces eq. (b)			
Multiply eq. (f) by 2	$2y =$	4	(g)
Subtract eq. (g) from eq. (c)	$x \quad =$	1	(h)

The solution can be found in equations (f) and (g), which have actually replaced the first two.

Large systems of linear equations frequently arise in scientific, engineering, and business applications. For example, problems in *operations research*, arising in management decisions concerning the best ways to produce and distribute products, may involve tens of thousands of equations in as many unknowns. We need a digital computer to solve them.

Suppose we are given the n equations

$$\begin{aligned} a_{11}x_1 + a_{12}x_2 + \ldots + a_{1n}x_n &= b_1 \\ a_{21}x_1 + a_{22}x_2 + \ldots + a_{2n}x_n &= b_2 \\ &\vdots \\ a_{n1}x_1 + a_{n2}x_2 + \ldots + a_{nn}x_n &= b_n \end{aligned} \tag{A9.1}$$

in the n unknowns $x_1, x_2, \ldots, x_n$, where $a_{11}, a_{12}, \ldots, a_{nn}$ and $b_1, b_2, \ldots, b_n$ are coefficients. We want to find values of $x_1, \ldots x_n$ that satisfy these equations. The method in hand computation would be to use the first equation to eliminate x_1 from the remaining $n - 1$ equations. This can be done by the following steps: Divide the first equation by a_{11} to get

$$x_1 + \frac{a_{12}}{a_{11}}x_2 + \ldots + \frac{a_{1n}}{a_{11}}x_n = \frac{b_1}{a_{11}} \tag{A9.2}$$

This is another linear equation equivalent to the original one, but with the coefficient of x_1 changed to 1. Next we subtract a_{21} times (A9.2) from the second of equations (A9.1) to get rid of x_1. We get

$$\begin{aligned} 0 \cdot x_1 + \left(a_{22} - a_{21}\frac{a_{12}}{a_{11}}\right)x_2 + \ldots + \left(a_{2n} - a_{21}\frac{a_{1n}}{a_{11}}\right)x_n \\ = \left(b_2 - a_{21}\frac{b_1}{a_{11}}\right) \end{aligned} \tag{A9.3}$$

This is another linear equation satisfied by any point $x_1, \ldots, x_n$ that satisfies the first two of equations (A9.1). Note that when we wrote the original equations (A9.1) and when we write the modified equations (A9.2) and (A9.3), we always write the form "coefficient times x_1 plus coefficient times x_2 plus ... plus coefficient times x_n equals number." The names of the unknowns $x_1, \ldots, x_n$, plus signs, and the equal sign always appear. Therefore there is no need to write them down. We need only write the coefficients a_{ij} and the numbers b_i. We just have to remember that the other signs are assumed to be there. Thus we can start with the rectangular array

$$\begin{array}{llllll} a_{11} & a_{12} & \cdots & a_{1n} & b_1 \\ a_{21} & a_{22} & \cdots & a_{2n} & b_2 \\ \cdot \\ \cdot \\ \cdot \\ a_{n1} & a_{n2} & \cdots & a_{nn} & b_n \end{array}$$

The first step is to divide the first row (that is, the first equation) by a_{11}. We call a_{11} the *pivot element*. This changes the first row so that the new element in the (1,1) position is one. We will call the element in the (i,j) position a_{ij} regardless of how many times it has been changed, because we are going to store the coefficients in an array A(I,J) when we solve this problem by computer. Hence the result of dividing the first row by the pivot element is

$$\begin{array}{llllll} 1 & a_{12} & \cdots & a_{1n} & b_1 \\ a_{21} & a_{22} & \cdots & a_{2n} & b_2 \\ \cdot \\ \cdot \\ \cdot \\ a_{n1} & a_{n2} & \cdots & a_{nn} & b_n \end{array}$$

The next step is to subtract a_{21} times the first row (equation) from the second row (equation) to reduce the first element of the second row to zero. This process is continued by subtracting a_{31} times the first row from the third row, and so on, until all elements below the pivot element in the first column are zero. The array now has the form

$$\begin{array}{llllll} 1 & a_{12} & \cdots & a_{1n} & b_1 \\ 0 & a_{22} & \cdots & a_{2n} & b_2 \\ \cdot \\ \cdot \\ \cdot \\ 0 & a_{n2} & \cdots & a_{nn} & b_n \end{array}$$

If we could find the solution for $x_2, x_3, \ldots, x_n$, the first row could be solved for x_1. The last $n - 1$ rows now represent $n - 1$ equations in the $n - 1$ unknowns $x_2, \ldots, x_n$. The same process can be applied to these, using a_{22} as the next pivot. The second row is divided by a_{22}, giving an array of the form

$$\begin{array}{llllll}
1 & a_{12} & a_{13} & \ldots & a_{1n} & b_1 \\
0 & 1 & a_{23} & \ldots & a_{2n} & b_2 \\
0 & a_{32} & a_{33} & \ldots & a_{3n} & b_3 \\
\cdot \\
\cdot \\
\cdot \\
0 & a_{n2} & a_{n3} & \ldots & a_{nn} & b_n
\end{array}$$

Now a_{32} times the second row is subtracted from the third row to put a zero in the (3,2) position. This is repeated with $a_{42}, \ldots, a_{n2}$ to put zeros in every element below the pivot a_{22} in the second column. The array now has the form

$$\begin{array}{llllll}
1 & a_{12} & a_{13} & \ldots & a_{1n} & b_1 \\
0 & 1 & a_{23} & \ldots & a_{2n} & b_2 \\
0 & 0 & a_{33} & \ldots & a_{3n} & b_3 \\
\cdot \\
\cdot \\
\cdot \\
0 & 0 & a_{n3} & \ldots & a_{nn} & b_n
\end{array}$$

Now the last $n - 2$ rows represent $n - 2$ equations in the $n - 2$ unknowns $x_3, \ldots, x_n$. Again we can repeat the process by making a_{33}, the new pivot, equal to one, and $a_{k3} = 0$ for $k = 4, 5, \ldots, n$. The process continues until all pivots a_{kk} are one, and all elements below the pivots are zero. Notice that the pivots are the diagonal elements of the array, and that all elements below the diagonal are zero. We say that the resulting array is *triangular*, because only a triangular part of it is nonzero, as shown in Figure A9.1.

The process we have been describing is called *Gaussian elimination*, after the mathematician Karl Friedrich Gauss. In the triangular form, the last row represents the equation $x_n = b_n$. Thus we have the solution

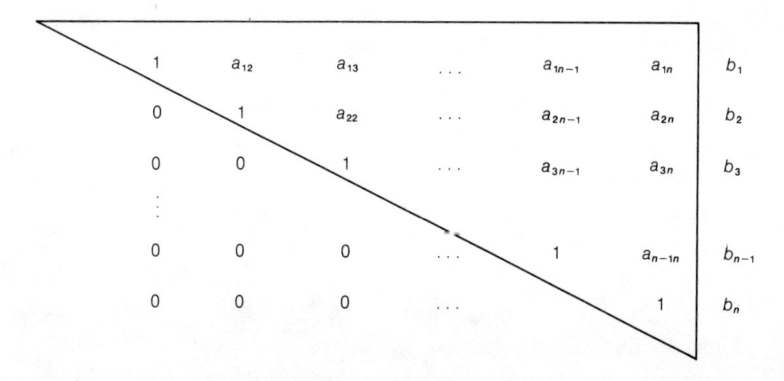

Figure A9.1 Triangular array

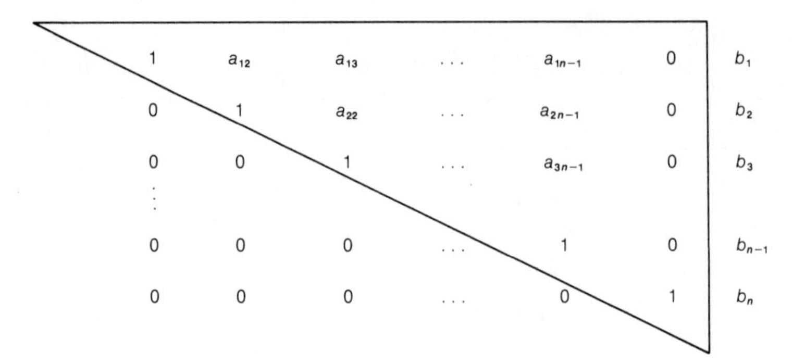

Figure A9.2 Array after back-substitution

for x_n. With this we can eliminate x_n from each of the other equations by subtracting a_{kn} times the last equation from the kth equation for $k = 1, 2, \ldots, n - 1$. The result is shown in Figure A9.2. Now the last row but one tells us that $x_{n-1} = b_{n-1}$. Next we can subtract suitable multiples of that row from the earlier rows to put zeros above the diagonal in the last column but one. This step can be repeated working back up the matrix until all the elements above the diagonal are zero. Now we have a *diagonal* array and the ith equation reads $x_i = b_i$. Thus the solution to the problem is found in the vector B. This second stage of the process is called *back-substitution*.

The implementation of this method in Program A9.1 uses a single array A of N by N + 1 elements to store the coefficients a_{ij} and b_i. The elements a_{ij} are stored in A(I,J) in the usual way. The elements b_i are stored in A(I,N+1), the last column of the array. This is convenient

Program A9.1 ***Outline of program to solve linear equations***

```
GAUSS: subprogram (A,N)
    Solve the system of N linear equations specified by the coefficients
    in the N-by-(N+1) array A. The Ith equation is

            A(I,1)*X1+A(I,2)*X2+ . . . +A(I,N)*XN=A(I,N+1)

    On completion, the solution X1, X2, . . . , XN is in the last column
    of A, elements A(1,N+1), A(2,N+1), . . . , A(N,N+1). The ele-
    ments of A are changed by the subprogram.
    real A(N,N+1),P,E
    integer I,J,K,N
        convert A to triangular form
        back-substitute
        return
    endsubprogram GAUSS
```

because a study of the algorithm reveals that the operations on the element b_i are identical to the operations on every element of the ith row of a_{ij}. Program A9.1 is presented at successively more detailed levels of refinement. The outline consists of two main segments: conversion of A to triangular form, which is given in successive levels of refinement in Programs A9.1a through A9.1c, and back-substitution, which is given in Program A9.1d. We have used the keywords codesegment and endcodesegment in Programs A9.1a through A9.1d because we intend these segments to be substituted back into the original outline. (They could be defined as subprograms and called in the outline, but the final program will not be long enough to warrant the extra computer time taken by procedure calls.)

Program A9.1a gives the code segment to triangularize the array in outline form. This segment is refined further in the next two programs. The division of row I by A(I,I) is shown in Program A9.1b. This code does not bother to divide elements to the left of the diagonal element A(I,I), because they are known to be zero. Neither does it bother to divide the diagonal element itself, because the answer is known to be one. There is no need to assign values to variables known to be zero or one, they are not used again in the computation.

Program A9.1c gives the code to subtract A(J,I) times row I from row J. Since the first I − 1 elements in both rows are known to be zero,

Program A9.1a ***Code segment to convert*** A ***to triangular form***

```
convert A to triangular form: codesegment
    The array A is modified by row operations consisting of division of
    all elements of one row by a common value, and subtraction of a
    multiple of one row from another.
        do for I←1 to N
            divide row I by A(I,I)
            do for J←I+1 to N
                subtract a multiple of row I from row J so that A(J,I)=0
                enddo
            enddo
    endcodesegment
```

Program A9.1b ***Code segment to divide row by pivot element***

```
divide row I by A(I,I): codesegment
    do for J←I+1 to N+1
        A(I,J)←A(I,J) / A(I,I)
        enddo
    endcodesegment
```

Program A9.1c ***Code segment to eliminate element*** A(J,I)

```
subtract a multiple of row I from row J . . . : codesegment
    do for K←I+1 to N+1
        A(J,K)←A(J,K)−A(I,K)*A(J,I)
        enddo
    endcodesegment
```

there is again no need to perform arithmetic on them. Since the result for A(J,I) is known to be zero, there is no point in computing or storing it either.

It is numerically preferable to change the order of operations slightly in the back-substitution segment (Program A9.1d). After the triangularization, we know that the value of x_n is in A(N,N+1). Hence we can write

$$x_{n-1} = b_{n-1} - x_n * a_{n-1,n}$$

which can be implemented by

$$A(N-1,N+1) \leftarrow A(N-1,N+1) - A(N,N+1)*A(N-1,N)$$

leaving the result in A(N−1,N+1). After we have calculated the values $x_n, x_{n-1}, \ldots, x_{j+1}$ in the elements A(K,N+1), for K from N to J + 1, we can calculate x_j by

$$x_j = -(x_{j+1} * a_{j,j+1} + x_{j+2} * a_{j,j+2} + \ldots + x_n * a_{j,n}) + b_j$$

In Program A9.1d, the term in parentheses is accumulated in the variable E.

Before we bring the pieces of this program together into a unit, there are two factors that should be considered: speed and accuracy. In Program A9.1b each element in a row is divided by the element A(I,I).

Program A9.1d ***Code segment to do back-substitution***

```
back-substitution: codesegment
    Outer loop to compute each x_j
    outer: do for J←N−1 to 1 by −1
        | E←0.0
        | Inner loop to form sum of known a_jk x_k
        | inner: do for K←J+1 to N
        |    | E←E+A(K,N+1)*A(J,K)
        |    | enddo inner
        | A(J,N+1)←A(J,N+1)−E
        | enddo outer
    endcodesegment
```

Execution of the statement in the third line of that segment requires that the address of the cell containing A(I,I) be calculated by the computer in order to read the value of the divisor from memory. Since I does not change inside the loop, it is not necessary to calculate the address of A(I,I) on every pass through the loop. Indeed, many compilers will recognize this fact and produce object code that is "optimized." However, if the compiler does not do this optimization, the programmer can do it by rewriting the loop as

```
P←A(I,I)
do for J←I+1 to N+1
    A(I,J)←A(I,J) / P
    enddo
```

Similarly, in Program A9.1c, the element A(J,I) does not change inside the loop, so the loop can be rewritten as

```
E←A(J,I)
do for K←I+1 to N+1
    A(J,K)←A(J,K)−A(I,K)*E
    enddo
```

Numerical accuracy in this type of process will not be considered until the next chapter. However, there is one situation in which the code shown clearly will not work: if the value of A(I,I) is zero for some I in Program A9.1b, the program will fail on a "divide by zero" error. Suppose that A(I,I) is zero when the time comes for this division. Notice that at this stage of the process, the rows I, I + 1, I + 2, . . . , N are similar, in that each has zeros in columns 1, 2, . . . , I − 1. Clearly, the solution of a system of linear equations is not affected by the order in which we write them so let us simply switch two of the equations, say rows I and K. If A(K,I) is nonzero before the switch, A(I,I) will be nonzero after the switch. Therefore, before dividing row I by A(I,I), we will make sure that A(I,I) is nonzero by switching rows if necessary. This will be possible if any one of the elements A(K,I), for K from I to N, is nonzero. (If all are zero, the problem does not have a well-defined solution.) It is convenient for reasons of accuracy (see Chapter A10) to switch rows so that the largest of the A(K,I) elements appears in position A(I,I). Program A9.1e is a modification of A9.1b that incorporates this change. The complete Gaussian-elimination subprogram is shown in Program A9.2.

Program A9.1e ***Improved version of Program A9.1b***

```
divide row I by A(I,I): codesegment
    Find largest A(K,I)
    K←I
```

```
P←ABS(A(I,I))
do for J←I+1 to N
  | if ABS(A(J,I))>P
  |     then
  |         K←J
  |         P←ABS(A(J,I))
  |     endif
  | enddo
if P=0.0
  | then
  |     output 'NO WELL DEFINED SOLUTION, CALCULATION
  |       TERMINATED'
  |     return
  | endif
```

If K≠I, switch rows I and K.

```
if K≠I
  | then
  |     do for J←I to N+1
  |       | P←A(I,J)
  |       | A(I,J)←A(K,J)
  |       | A(K,J)←P
  |       | enddo
  | endif
```

Now divide by the pivot element

```
P←A(I,I)
do for J←I+1 to N+1
  | A(I,J)←A(I,J) / P
  | enddo
endcodesegment
```

Program A9.2 ***Solve set of linear equations by Gaussian elimination***

```
GAUSS: subprogram (A,N)
```

The N-by-(N+1) array A contains the coefficients of a set of N linear equations. The Ith equation is

```
A(I,1)*X1+A(I,2)*X2+ . . . +A(I,N)*XN=A(I,N+1)
```

The subprogram leaves the solution x_i in element A(I,N+1). If no solution is possible, the value of GAUSS is 0, otherwise 1.

```
integer GAUSS,I,J,K,N
real A(N,N+1),P,E
```

Convert A to triangular form

```
do for I←1 to N
```

```
    Find largest A(K,I)
    K←I
    P←ABS(A(I,I))
    do for J←I+1 to N
        if ABS(A(J,I))>P
            then
                K←J
                P←ABS(A(J,I))
            endif
        enddo
    if P=0.0
        then
            output 'NO WELL DEFINED SOLUTION,
              CALCULATION TERMINATED'
            return (0)
        endif
    If K≠I, switch rows I and K.
    if K≠I
        then
            do for J←I to N+1
                P←A(I,J)
                A(I,J)←A(K,J)
                A(K,J)←P
                enddo
        endif
    Divide row I by pivot element A(I,I)
    P←A(I,I)
    do for J←I+1 to N+1
        A(I,J)←A(I,J) / P
        enddo
    Eliminate A(J,I) for J from I + 1 to N by subtracting A(J,I)
    times row I from row J.
    do for J←I+1 to N
        E←A(J,I)
        do for K←I+1 to N+1
            A(J,K)←A(J,K)−A(I,K)*E
            enddo
        enddo
    enddo
Back-substitute
Outer loop to compute each x_j
outer: do for J←N−1 to 1 by −1
    E←0.0
    Inner loop to form sum of known a_jk x_k
```

```
        inner: do for K←J+1 to N
            E←E+A(K,N+1)*A(J,K)
            enddo inner
        A(J,N+1)←A(J,N+1)−E
        enddo outer
    return (1)
endsubprogram GAUSS
```

Problems

1. Some systems of linear equations, called *sparse* equations, have large numbers of zero coefficients. If there are many zeros, special techniques (which are beyond the scope of this text) can be used to reduce the amount of storage and arithmetic performed in the solution of the system. If there are not many zeros, these techniques are not valuable, but some speed improvement can be obtained by testing to see whether some of the elements below the diagonal are already zero, and omitting the elimination step for the corresponding rows. Modify Program A9.2 to include such a test.

†**2.** Compute the number of times that each statement in Program A9.2 is executed for N = 3, 4, and 5. (Assume that the conditions in the if statements are always true except that P ≠ 0.0.) Can you give the number of times each statement is executed as a function of N? You will need the relations

$$1 + 2 + 3 + \ldots + N = N(N + 1)/2$$
$$1^2 + 2^2 + 3^2 + \ldots + N^2 = N(N + 1)(2N + 1)/6$$

3. Program A9.2 can fail to find a solution because of overflow. The most likely place for this to occur is in the step where each element in a row is divided by the pivot element. An alternative scheme is to leave this step for later. In that case, the elimination step must be changed by replacing the value of E with A(J,I) / A(I,I). (This cannot cause overflow—why?) Also, the back-substitution step must be changed to divide the value computed for x_j by A(J,J). This can cause an overflow, but a test can be programmed so that a return with GAUSS = 0 can be made if an overflow would occur. Modify the program in this way.
4. Program A9.2 swaps rows of the array A to put the largest element on the diagonal as a pivot. This process is time consuming. An alternative scheme is to use a one-dimensional array of pointers to indicate which row of A is used for the Ith pivot. Suppose this array is called ROW. Initially, ROW can be set so that ROW(I) contains I for I from 1 to N. If, when selecting the pivot in the Ith column, the Kth row is selected, the Ith and Kth elements of ROW can be exchanged.

Instead of swapping the Ith and Kth rows, we can reference the Ith row by A(ROW(I),K) for K from 1 to N. If all references to rows of A are made through the array ROW in this way, Program A9.2 will compute the solutions so that x_i is in A(ROW(I),N+1).

a. Modify Program A9.2 in the manner described.

*b. Design a scheme to unscramble the answers, so that the solution for x_i is in A(I,N+1). (Hint: Scan through ROW, and if ROW(I) is not I, move A(ROW(I),N+1) to A(I,N+1), saving the old value of A(I,N+1). Before continuing the scan, check ROW(ROW(I)) and find out what should be stored in A(ROW(I),N+1). Continue in this way until you get back to the Ith element, then continue the scan.)

Chapter A10

Numerical Error

This chapter takes a brief look at *numerical error.* Such errors initially arise from two sources. The first source is called *rounding error*, and is due to the fact that only a finite subset of all the real numbers can be represented in the computer as floating-point numbers. Therefore we must approximate most input data and the results of most arithmetic operations. The second source of numerical error is called *truncation error.* It has nothing to do with the use of truncation as a means of converting real numbers to representable floating-point numbers, but refers to the approximation of functions that can be computed exactly only by an infinite sequence of operations. For example, sin(y) is given by the series

$$y - \frac{y^3}{3!} + \frac{y^5}{5!} - \frac{y^7}{7!} + \ldots$$

If a finite number of terms of this series are used to compute sin(y), the error due to truncating the series is called the truncation error. The errors introduced by rounding and truncation are initially small, but sometimes their effect is amplified by subsequent operations.

The purpose of this chapter is simply to point out some of the pitfalls in numerical computation, not to discuss numerical methods exhaustively. Even if you are not particularly concerned with numerical computations, you should still be aware of the traps awaiting the unwary.

A10.1 ERRORS IN ARITHMETIC [ROUNDING ERROR]

Arithmetic error arises when initial data or intermediate results cannot be represented exactly in the computer. Because machines are usually more precise than we are, we often assume that the computer is precise enough for anything we need to do. Our hypothetical memory carries five digits of precision in floating point. Since numbers we measure in experiments are seldom accurate to more than three digits, surely five are sufficient? Besides, most actual computers carry from seven to sixteen digits of precision. Surely that is enough, even if our experimental data is good to six digits? Often it is, but we must be aware that things can go wrong.

The number of digits of precision carried in floating point is fixed. Our hypothetical computer has five. The *weight* of these digits varies with the exponent of the number. Thus if a number is rounded to put it in the form

$$\pm.nnnnn \times 10^e$$

the rounding error will be a maximum of $\pm 0.000005 \times 10^e$. Its actual value will depend on the exponent e. When an arithmetic operation results in an answer that cannot be represented exactly, additional rounding errors are introduced. These depend on the exponent of the answer.

Example A10.1 *Rounding Errors*

A common example of the buildup of rounding errors is in the repeated addition of the same number to a running total. Suppose that in the nth pass through a program loop we need the value $n/6$. One way of arranging this is to initialize a variable to 1/6 and add 1/6 to it after each pass. In order to see what happens after only a few passes, let us consider the behavior of a machine with only two significant digits of precision in floating point. One-sixth is represented by 0.17×10^0 with rounding, by 0.16×10^0 with truncation.

Let us look at the state of the variable in the first twelve passes in each case. They are shown in Table A10.1. Once the total reaches 1.0, future additions of 1/6 can only add 0.1 or 0.2, because the intermediate result has an exponent of 1. If the passes through the loop continued until the total were 10, further additions would not change the result, since $10 + 0.16$ is still 10 to two significant digits. It is true that modern computers have more than two significant digits, but it is also true that we often make more than twelve passes through a loop.

One way to overcome this phenomenon is to form an integer N that is equal to the number of passes through the loop and divide that by 6. It takes a little more computer time, since division is slower than addition, but it is much more accurate, because the maximum error is a rounding error in the last significant digit of the answer. This example demonstrates that the fastest program may not be the most accurate.

TABLE A10.1 RUNNING SUM OF 1/6

Pass	Sum Correctly Rounded	Sum Truncated	True Answer Rounded
1	$.17 \times 10^0$	$.16 \times 10^0$	$.17 \times 10^0$
2	$.34 \times 10^0$	$.32 \times 10^0$	$.33 \times 10^0$
3	$.51 \times 10^0$	$.48 \times 10^0$	$.50 \times 10^0$
4	$.68 \times 10^0$	$.64 \times 10^0$	$.67 \times 10^0$
5	$.85 \times 10^0$	$.80 \times 10^0$	$.83 \times 10^0$
6	$.10 \times 10^1$	$.96 \times 10^0$	$.10 \times 10^1$
7	$.12 \times 10^1$	$.11 \times 10^1$	$.12 \times 10^1$
8	$.14 \times 10^1$	$.12 \times 10^1$	$.13 \times 10^1$
9	$.16 \times 10^1$	$.13 \times 10^1$	$.15 \times 10^1$
10	$.18 \times 10^1$	$.14 \times 10^1$	$.17 \times 10^1$
11	$.20 \times 10^1$	$.15 \times 10^1$	$.18 \times 10^1$
12	$.22 \times 10^1$	$.16 \times 10^1$	$.20 \times 10^1$

As Example A10.1 shows, rounding errors can accumulate to the point that the answer is meaningless. If that were the only way rounding errors could become significant, the analysis of numerical error would be relatively easy. Unfortunately, if we allow intermediate results to become large, they will include large rounding errors. If the errors remain the same size when they are transmitted to the solution, the answer will be lost in the error.

Large rounding errors are transmitted to the answer when two large numbers of almost the same size are subtracted. This phenomenon is known as *cancellation*. Consider, for example, the problem of subtracting 129/388 from 162/485. The correct answer is 3/1940. If the numbers are correctly rounded to five significant digits, the arithmetic process yields

$$0.33402 \times 10^0 - 0.33247 \times 10^0 = 0.15500 \times 10^{-2}$$

whereas the answer, correctly rounded to five digits, is 0.15464×10^{-2}. (See Figure A10.1.) What has happened? When the fractions were represented to five-digit precision, rounding errors were introduced. The true value of 129/388 is 0.3324742 . . . , so a rounding error of 0.0000042 . . . $\times 10^0$ is committed. Similarly, a rounding error of 0.0000006 . . . $\times 10^0$ is made in representing 162/485 to five digits. When these two numbers are subtracted, their first two digits cancel, decreasing the exponent of the result by 2. The total rounding error is $(0.0000006 - 0.0000042) \times 10^0$, which is $(-0.00036) \times 10^{-2}$, and so it shows up as early as the fourth place of the answer. However, the error is not *caused* by cancellation; it is only uncovered by cancellation. The error occurred when the numbers were rounded. At that time the error was in the sixth significant place; cancellation moved it to a more significant position.

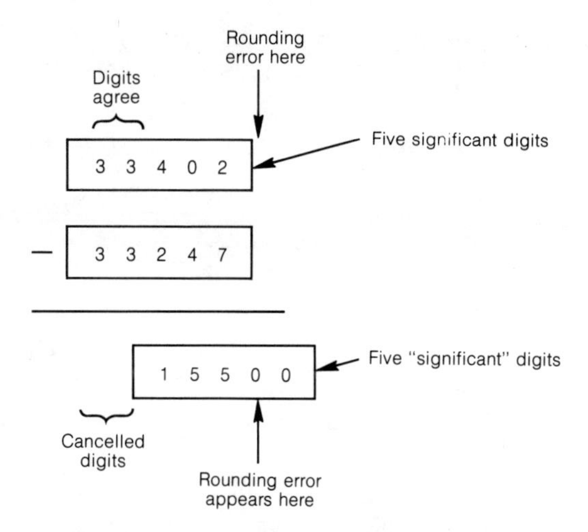

Figure A10.1 Cancellation

A well-known example of cancellation is the addition of three numbers. Suppose we form A+B+C in five-digit floating-point arithmetic by first adding A to B, then adding C to the result. Suppose the values of A, B, and C are -0.12344×10^0, $+\ 0.12345 \times 10^0$, and $+0.32741 \times 10^{-4}$, respectively. A+B is 0.00001×10^0, or 0.10000×10^{-4} after normalization. When C is added to this, we get 0.42741×10^{-4}. Suppose, on the other hand, we first add C to B, then add the result to A. B+C is 0.123482741×10^0, which is 0.12348×10^0 after rounding. When A is added to this, we get 0.00004×10^0, which is 0.40000×10^{-4} after normalization. Thus the result (0.42741×10^{-4} or 0.40000×10^{-4}) is dependent on the order of the arithmetic. Although exact addition is *associative*—that is, (A+B)+C = A+(B+C)—the result of machine computation may not be.

Example A10.2 *Variance Calculation*

If the N numbers in a set are S(1), S(2), . . . S(N), then their *mean* (or *average*) is

$$A = (S(1) + S(2) + \ldots + S(N))/N$$

and their *variance* is

$$V = ((S(1) - A)^2 + (S(2) - A)^2 + \ldots + (S(N) - A)^2)/N$$

The variance is a measure of the scattering of a set of values. If all the values are the same, the variance is zero. The mean and variance can be calculated from the formulas above using two loops, the first to calculate the mean by summing the values, and the second to calculate the variance by summing $(S(I) - A)^2$. This takes 2N − 2 additions, N subtractions, N multiplications, and 2 divisions. Alternatively, we can rewrite V as

$$V = ((S(1)^2 - 2S(1) \times A + A^2) + \ldots + (S(N)^2 - 2S(N) \times A + A^2))/N$$
$$= (S(1)^2 + S(2)^2 + \ldots + S(N)^2)/N - 2 \times A \times (S(1) + S(2) + \ldots + S(N))/N + (A^2 + A^2 + \ldots + A^2)/N$$

The second term contains $(S(1) + S(2) + \ldots + S(N))/N$ which is exactly A, while the last contains N instances of A^2. Hence

$$V = (S(1)^2 + S(2)^2 + \ldots + S(N)^2)/N - 2 \times A \times A + A^2$$
$$= (S(1)^2 + S(2)^2 + \ldots + S(N)^2)/N - A^2$$

A program to compute this is shown in Program A10.1.

This approach requires N − 1 fewer subtractions and only one more multiplication than the earlier method, and uses only one loop, resulting in a faster program. However, if the average (A) is large but the variance (V) is small, this faster method may be much less accurate. Consider an example using a three-digit computer that rounds correctly. Suppose we wish to find the variance of the five numbers 4.3, 4.4, 4.5, 4.6, and 4.7. Using the earlier method, we first sum them to get 22.5. On dividing by 5 we get the correct average of 4.5. No rounding errors are introduced if three digits are used. Next we calculate the sum of the squares of the differences, as shown in Table A10.2. The computed variance is

Program A10.1 ***Calculate variance the fast but inaccurate way***

```
VAR: subprogram (S,N)
    The value of this function is the variance of the N values in the real
    array S.
    integer I,N
    real VAR,S(N),V,A
        V←0.0
        A←0.0
        do for I←1 to N
            Form sum in A and sum of squares in V.
            A←A+S(I)
            V←V+S(I)↑2
            enddo
        A←A/FLOAT(N)
        V←V/FLOAT(N)−A↑2
        return (V)
    endsubprogram VAR
```

TABLE A10.2 VARIANCE CALCULATION BY SLOW METHOD

$(4.3 - 4.5)^2$	$= 0.4 \times 10^{-1}$
$(4.4 - 4.5)^2$	$= 0.1 \times 10^{-1}$
$(4.5 - 4.5)^2$	$= 0.0$
$(4.6 - 4.5)^2$	$= 0.1 \times 10^{-1}$
$(4.7 - 4.5)^2$	$= 0.4 \times 10^{-1}$
Total	$= 0.1 \times 10^{0}$
Variance = Total/5	$= 0.2 \times 10^{-1}$

$0.1 \times 10^0/5 = 0.2 \times 10^{-1}$. Again no rounding errors are committed, so the answer is exact. By contrast, the second method will square each of the S(I)'s and round the answers to three digits. This process and the total are shown in Table A10.3. This time we calculate V using three-digit rounded arithmetic, to get

$$V = \frac{102.}{5} - (4.5)^2 = 20.4 - 20.3 = 0.1$$

This is in error by 400%! An extreme example, yes—but one to remind you that the fastest way is not always the best.

A10.2 TRUNCATION ERROR

Truncation error arises because an infinite process necessary to compute a function must be truncated after a finite number of steps. We have already mentioned the example of $\sin(y)$. Many of the simple functions are given by power series. For example

$$\exp(y) = 1 + y + \frac{y^2}{2!} + \frac{y^3}{3!} + \ . \ . \ .$$

If we want to know how many terms to use, we must decide how much precision we need, and the range of y to be allowed. If, for example, we are interested in $-1 \leq y \leq 1$, the difference between $\exp(y)$

TABLE A10.3 VARIANCE CALCULATION BY FAST METHOD

I	S(I)	S(I)²	Rounded to Three Digits	Sum So Far, Rounded to Three Digits
1	4.3	18.49	18.5	$18.5 = S(1)^2$
2	4.4	19.36	19.4	$37.9 = S(1)^2 + S(2)^2$
3	4.5	20.25	20.3	$58.2 = S(1)^2 + S(2)^2 + S(3)^2$
4	4.6	21.16	21.2	79.4 = . . .
5	4.7	22.09	22.1	102. = . . .
Variance = Sum/5 − 20.3 = 0.1×10^0				

TABLE A10.4 VALUES OF MAXIMUM ERROR IN APPROXIMATION TO exp(y)

n	2	3	4	5
E_n	0.218 . . .	0.0516 . . .	0.00995 . . .	0.00161 . . .

and $1 + y + \ldots + y^n/n!$ is certainly less than

$$E_n = \frac{1}{(n+1)!} + \frac{1}{(n+2)!} + \ldots$$

Since

$$1 + 1 + \frac{1}{2!} + \frac{1}{3!} + \ldots = e = 2.7182818\ldots$$

we can calculate the maximum error for various n to be

$$E_n = 2.7182818\ldots - \left(1 + 1 + \frac{1}{2!} + \ldots + \frac{1}{n!}\right)$$

Values of E_n are shown in Table A10.4. A person designing a function procedure for exp could pick a value of n from a table such as this. (Not all functions can be treated this simply.)

Even if computer arithmetic were infinitely accurate, stopping the calculation of a series after a finite number of terms leads to truncation error. This type of error also occurs in some of the other examples we have discussed. The method of bisection, for example, computes more accurate approximations to the solution of an equation by moving the upper and lower bounds closer together. But no matter how much accuracy is achieved in the arithmetic or how many steps are taken, the two bounds will always be separate, so we will not get an exact answer unless we are lucky and happen to hit the answer when one of the midpoints is formed.

A10.3 AMPLIFICATION OF ERRORS

Most problems start with data that contains errors. These may come, for example, from incorrect measurement in experiments. Obviously, errors in the data will lead to errors in the answers. Rounding and truncation error also lead to errors in the answers. Even though all such errors are of limited size—for example, less than 0.1%—it does not follow that the errors in the final answer will be similarly restricted in size. Indeed, in the examples above, the final errors were much larger, because of large intermediate results. Even in methods of solution that involve no truncation error, and in which arithmetic is done with arbitrary accuracy, errors in the initial data can be magnified out of all proportion.

Such problems are said to be *ill-conditioned.* In an ill-conditioned problem, no method can lead to an accurate answer, so we cannot come up with a solution to the difficulty. However, it is important to know when ill-conditioning can arise, so that we will know when our answers are nonsense.

An example of an ill-conditioned problem is that of deciding where a moon-bound rocket is going to land, assuming that it is aimed from the earth and that no midcourse corrections are allowed. The effect on the final destination of errors in the initial angle and velocity is critical, since a small variation will leave the rocket wandering in space or crashing into the sun, 9×10^7 miles off course!

In this section we will start with a simple example of an ill-conditioned problem. We will then investigate a method of solution of a problem that is initially well-conditioned but can be changed into an ill-conditioned problem if care is not exercised. In this case a good problem is ruined by a poor choice in the method of solution.

Consider the following system of two equations in two unknowns:

$$\begin{aligned} 0.992u + 0.873v &= 0.119 \\ 0.481u + 0.421v &= 0.060 \end{aligned} \tag{A10.1}$$

These can be seen by inspection to have the solution $u = 1$, $v = -1$. Suppose these numbers are the result of measurements in an experiment, and are good only to within ± 0.001. Thus the right-hand side of the first equation might be 0.120 rather than 0.119. Let us solve the *perturbed* problem

$$\begin{aligned} 0.992u + 0.873v &= 0.120 \\ 0.481u + 0.421v &= 0.060 \end{aligned} \tag{A10.2}$$

The answers are

$$\begin{aligned} u &= 0.815 \\ v &= -0.789 \end{aligned}$$

to three digits, about a 20% change caused by a 1% change in the problem. It is easy to construct examples in which the change in the answer is arbitrarily as large. For example

$$\begin{aligned} 0.400y + 0.400z &= 0.800 \\ 0.401y + 0.400z &= 0.801 \end{aligned} \tag{A10.3}$$

has the solution $y = z = 1$, whereas

$$\begin{aligned} 0.400y + 0.400z &= 0.800 \\ 0.401y + 0.400z &= 0.800 \end{aligned} \tag{A10.4}$$

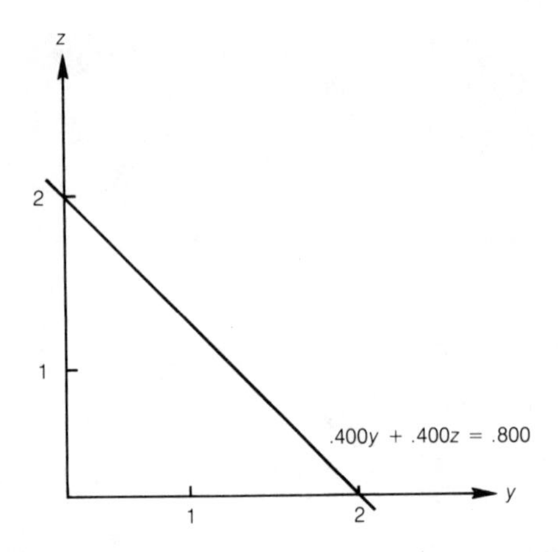

Figure A10.2 Graph of $0.400y + 0.400z = 0.800$

has the solution $y = 0$, $z = 2$. A change of one part in 800 gives a 100% change in the answer!

What is happening in these examples? A pictorial representation makes it clear. The set of values of y and z for which

$$0.400y + 0.400z = 0.800$$

form a line on a graph of y versus z, as shown in Figure A10.2. This line represents all those points (y,z) that satisfy the first of equations (A10.3). If we recognize that the initial data has errors, so that the equation might really be $0.400y + 0.400z = 0.801$, we see that points on this perturbed line could equally well satisfy the first equation *within the accuracy we were able to measure*. When we consider all possible equations that could replace the first one if we allow each of the coefficients to be perturbed by any amount up to its maximum error, we see that the set of points that satisfy the first equation (or rather, satisfy what we know about it) forms a region like the "thick line" shown in Figure A10.3. (Actually, the thickness of that line is exaggerated: it is only about 0.004 units thick.)

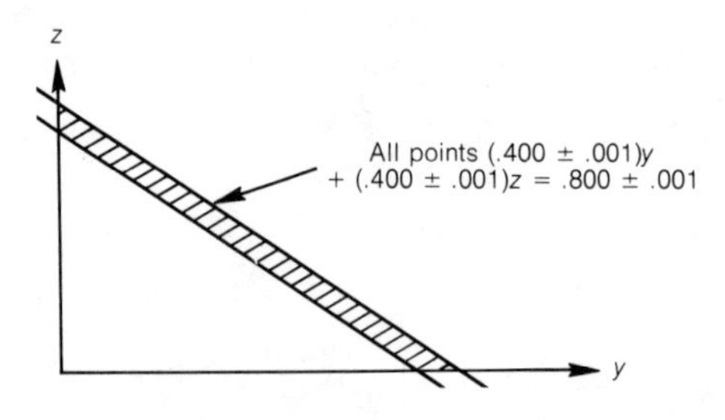

Figure A10.3 Region of solution

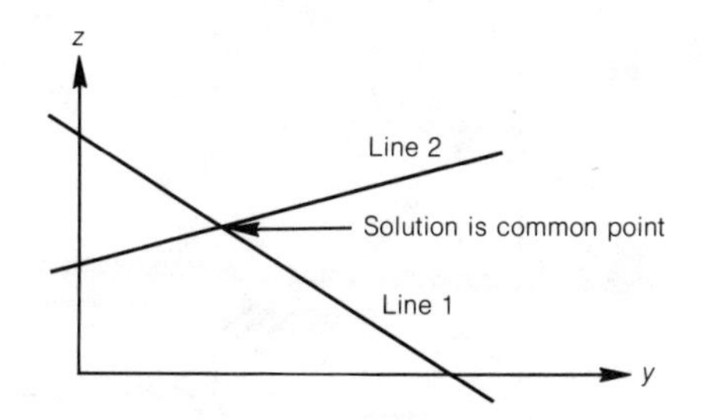

Figure A10.4 Solution of a pair of equations

The second equation also represents a line if the coefficients are known exactly (see Figure A10.4), or a "thick line" if there are errors in the coefficients. The solution of a pair of equations, each of which is known exactly, is the point of intersection of two lines, as shown in Figure A10.4. When we allow for error, a possible solution will be any point in the common region as shown in Figure A10.5. It is not possible to determine y and z more precisely than shown if the given data is approximate. This is all right if the lines are only about 0.004 thick and oriented as shown. The uncertainty in y and z is only about 0.007. However, suppose the two lines are nearly parallel, as shown in Figure A10.6. In that case, the common region is very large, even though the lines are not very thick—that is, even though the initial data is fairly accurate. The examples we gave above both represented nearly parallel lines. Nothing can be done to determine the answer more accurately than it is determined by the original problem. If the answer is very sensitive to changes in the original data, the problem is ill-conditioned.

However, we can hope to determine a solution that lies in the region common to the two "thick lines," or at least close to both of the original lines. That is to say, we should be able to solve a problem that is no more than a small peturbation of the original problem. If we do and the problem is well-conditioned, the computed solution will be close to the cor-

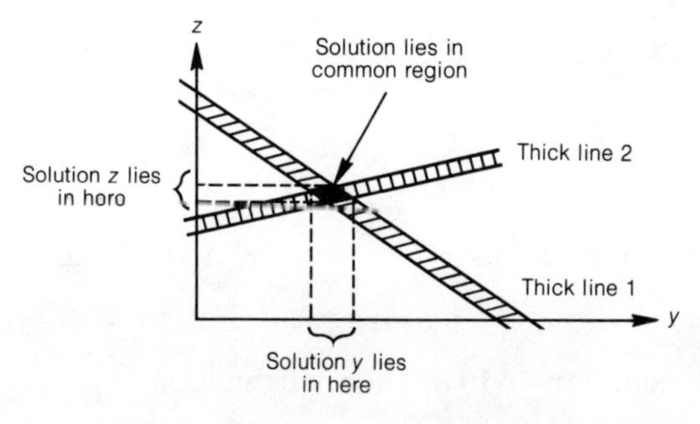

Figure A10.5 Area of possible solution

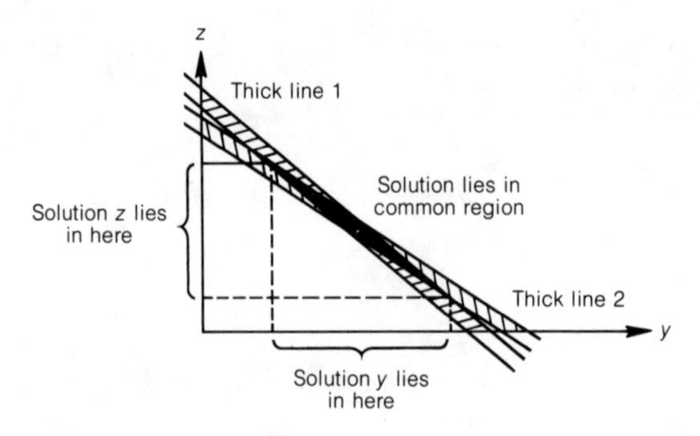

Figure A10.6 Area of possible solution with nearly parallel lines

rect solution. If the problem is ill-conditioned, the computed solution may not be close to the correct solution, but it will be no worse than the solution to a "nearby" problem—and since the original data may well have been in error, the solution to the real problem could be equally far from the solution of the problem solved as is the computed solution.

If we are given a well-conditioned problem, we should be able to determine the answers accurately; but sometimes a bad choice of method will make the solution very sensitive to rounding or truncation error, or even to errors in the initial values. This is particularly true in solving linear equations. Let us consider an example using three-digit floating-point arithmetic.

$$\begin{aligned} 0.512y - 0.920 \times 10^{-3}z &= 0.511 \\ 0.117 \times 10^{-2}y + 0.648z &= 0.649 \end{aligned} \qquad \text{(A10.5)}$$

The two lines represented are almost at right angles, so the problem is well-conditioned. (The answer is $y = z = 1$ to three digits.) The usual method of solution is to use the first equation to express the first variable (y) in terms of the other variables. In this case we divide the first equation by 0.512 to get

$$y - 0.180 \times 10^{-2}z = 0.998 \qquad \text{(A10.6)}$$

and then subtract 0.117×10^{-2} times equation (A10.6) from the second of equations (A10.5) to get y. Thus we have

$$\begin{matrix} 0.117 \times 10^{-2}y + 0.648z \\ -0.117 \times 10^{-2}(y - 0.180 \times 10^{-2}z) \end{matrix} = \begin{matrix} 0.649 \\ -0.117 \times 10^{-2} \times 0.998 \end{matrix}$$

If we do this arithmetic rounded to three digits we get

$$0.648z = 0.648$$

or $z = 1.00$. Equation (A10.6) then tells us that

$$y = 0.998 + 0.180 \times 10^{-2} = 1.00$$

to three digits. Suppose, however, that we had written equations (A10.5) in the reverse order:

$$\begin{aligned} 0.117 \times 10^{-2}y + 0.648z &= 0.649 \\ 0.512y - 0.920 \times 10^{-3}z &= 0.511 \end{aligned} \qquad \text{(A10.7)}$$

We divide the first of equations (A10.7) by 0.117×10^{-2} to get

$$y + 554z = 555 \qquad \text{(A10.8)}$$

Now we subtract 0.512 times equation (A10.8) from the second of equations (A10.7) to get rid of y:

$$\begin{aligned} 0.512y - 0.920 \times 10^{-3}z &= 0.511 \\ -0.512(y + 554z) &= -0.512 \times 555 \end{aligned}$$

In three-digit rounded arithmetic we get

$$-284z = -283$$

or

$$z = 0.996$$

Substituting this into equation (A10.8) we find

$$y = 555 - 554 \times 0.996 = 555 - 552 = 3.00$$

The answer for z is reasonable (0.4% error, which is all we can expect in view of the initial error). The answer for y is hopeless, and yet the problem is well-conditioned. What has happened? The answer is that we have turned it into an ill-conditioned problem in the middle of the solution by making a bad choice of which equation to handle first. Small rounding errors or initial errors then perturb the solution of this new ill-conditioned problem. Again we can see what is happening if we use a graphical interpretation.

When we divide an equation to make the coefficient of one of the variables equal to one, we do not change the line it represents, so this step has little effect. When we subtract a multiple of one equation from another, we get a third equation, which also represents a straight line. Any values of y and z that lie on both of the original lines must also lie on this third line, since if both of the original equations are satisfied, so is the third. Hence this line passes through the intersection of the other two, as shown in Figure A10.7. We can replace the problem of finding the common point of line 1 and line 2 by that of finding the common point of lines 1 and 3. We chose line 3 so that y did not appear in its equation: hence it is parallel to the y axis. In Figure A10.7, line 1 is nearly vertical, so our new problem appears to be well-conditioned.

In equations (A10.5), the first equation corresponds to a nearly vertical line, the second to a nearly horizontal line. When we eliminate

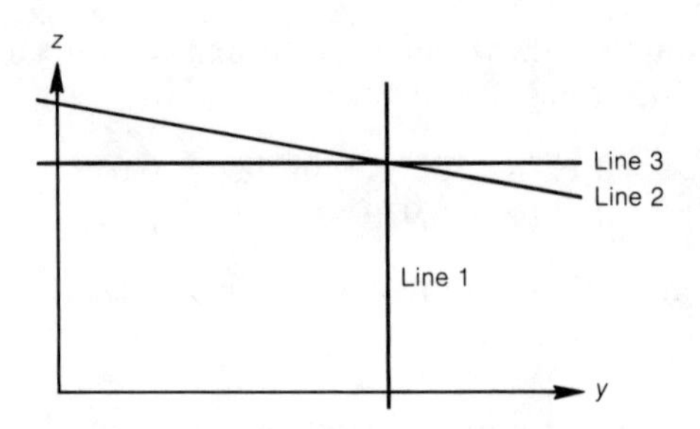

Figure A10.7 Effect of elimination of variable

y from the second equation to get a horizontal line, we still have lines nearly at right angles, so we still have a well-conditioned problem. However, when we reversed the order and kept the second and third lines by eliminating the first equation, we gave ourselves an ill-conditioned problem. This is shown with "thick lines" in Figure A10.8. The original region of uncertainty is $ABCD$. Line 3 must contain this region, since any solution of the original problem must lie in the thick line 3. If lines

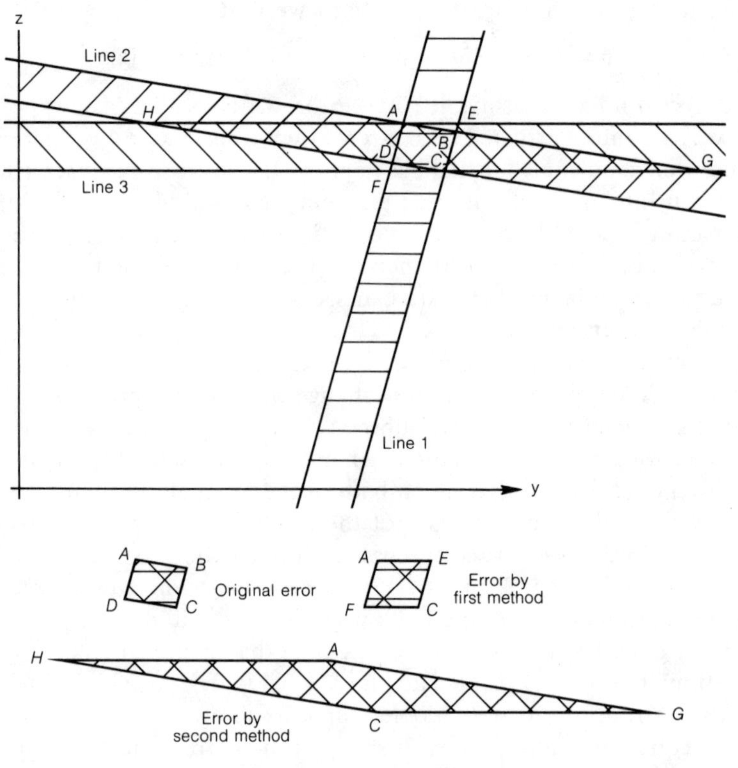

Figure A10.8 Regions of uncertainty

1 and 3 are now used to find the solution, the minimum region of uncertainty is *AECF*; if lines 2 and 3 are used, the region is much larger, namely *AGCH*.

In a large system of equations, similar problems can arise. One indication of potential trouble occurs when a pivot element is small. Referring back to Program A9.1a, we see that the diagonal element A(I,I) is selected as the pivot and will be used to eliminate all elements below A(I,I) in column I. If A(I,I) is small compared to an element to be eliminated, say A(J,I), then other elements in row J may be increased greatly when the program subtracts A(J,I) / A(I,I) times row I from row J. If the elements in row J become large, they will be subject to large rounding errors, which will later be revealed by cancellation. We can see this process in equation (A10.8). The coefficients have become large, so when the value of z is substituted, cancellation occurs, revealing the error.

It is difficult to avoid some amplification of error in this problem. The solution usually employed is *partial pivoting*. In this technique, the rows of the problem are exchanged as necessary to make A(I,I) the largest of the elements A(K,I), for K from I to N, before the Ith pivot is chosen. This process was implemented in Programs A9.1e and A9.2. A better process is called *full pivoting*. In full pivoting, the largest of the elements A(J,K), for J and K from I to N, is chosen as the pivot. A(J,K) can be moved into the (I,I) position by interchanging rows I and J and columns I and K. Interchanging rows has no effect; interchanging columns means that the variables are associated with different columns. Full pivoting is not often used because the resulting program is slower.

Problems

***1.** What is rounding error?

2. What is cancellation?

***3.** How can cancellation magnify the effect of rounding error in the final answer?

4. What is truncation error?

5. Compute the mean and variance of the numbers 4.9, 5.05, 4.98, 4.89, and 5.02 using correctly rounded three-digit arithmetic by the techniques used in Tables A10.2 and A10.3.

6. What is an ill-conditioned problem?

***7.** Is the problem of computing the hypotenuse of a right triangle, given the other two sides, well-conditioned?

8. Solve the equations

$$\begin{aligned} -0.263a + 0.527b &= 0.392 \\ 0.826a + 0.412b &= 0.851 \end{aligned}$$

by hand, using three-digit rounded floating-point arithmetic.

9. Solve the equations

$$0.215x + 0.430y = 0.258$$
$$0.461x + 0.817y = 0.553$$

by hand, using three-digit rounded floating-point arithmetic. Are your answers accurate? Is the problem well-conditioned?

10. a. Modify Program A9.2 to use full pivoting. You should use an array of N integers to record which column of A is used for the Ith pivot. If this array is COLUMN, it should be set initially so that COLUMN(I) = I for I from 1 to N. If, when the Ith pivot is being selected, the Kth column is used, the Ith and Kth elements in COLUMN should be switched. All columns of A should be referenced through COLUMN. That is, to find the Ith column of A, we write A(J,COLUMN(I)) for J from 1 to N.

*b. If this is done, how do we find the solution for x_i when the program has finished execution?

Chapter

A11 Simulation

Simulation is a very important tool in many areas of business planning, economics, engineering, and science. In this technique, a model of the real world is analyzed, usually by the computer. A *model* is a set of assumptions about the behavior of a real-world system. For example, it may describe the behavior of an electronic circuit or the expected arrival pattern of customers at a bank. The behavior of the model is studied to determine the effects of various changes without actually making those changes. There are two very different types of simulation in common use, *continuous* and *discrete.* Both forms are intended to predict what will happen to a given system as time passes. In continuous simulation, the model deals with variables (such as distance, voltage, force, etc.) that take on real values and change continuously with time. For example, a mathematical model of a building might be subjected to the forces of a typical earthquake to see the extent of the damage. (Simulation of this process is considerably easier and less expensive to plan than an actual experiment.) In discrete simulation, the model deals with *events* that occur at isolated times: for example, the arrivals of customers at a bank. The model might be used to determine the average waiting time for each customer under various organizations of tellers and lines. In this case, it might be feasible to experiment in a real-world situation; but in a study of, say, the national economy, it is probably better to simulate the effects of a measure such as a tax cut than to try it out in the real world and see what happens.

A significant part of the science (or art) of simulation is the design of an appropriate model. In business simulation, the planner tries to define models that are simple enough to make simulation practical, but comprehensive enough to model all the important effects. Part of the process of simulation is *verification*—comparing the results of simulated and actual experiments to see how well the model agrees with the real world. Only if these results show reasonable agreement can the planner have some degree of faith in the results of further simulations.

Similar considerations apply in science and engineering. If the purpose of a simulation is to design a piece of equipment without having to construct and try out each new design, the model used must be verified by comparing it against some experimental results. In science, simulation is also used to verify or disprove hypotheses. For example, a physicist may hypothesize that certain phenomena behave according to some mathematical equation (model); the simulation either disproves the hypothesis or verifies that it may be true. (It cannot *prove* that it is true!)

The techniques used in discrete and continuous simulation are totally different. The detailed mathematics of the former are part of numerical analysis, and the development of an appropriate model requires a basic understanding of the physics or economics of the problem. The details of discrete simulation are concerned with statistics and programming techniques.

Because simulation has been recognized as a valuable tool, there are many programming languages designed especially for simulation applications. When a problem is described in such a *problem-oriented language*, the method of solution does not have to be specified, because it is built into the language processor. GPSS (General Purpose System Simulation), for example, allows a user to describe a discrete-system model. GPSS interprets the description of the model and simulates its behavior. Another simulation language, ECAP (Electronic Circuit Analysis Program), accepts a description of an electronic circuit and determines its behavior. Thanks to such special-purpose languages, most users of simulation do not need to understand how the simulator itself operates. However, since somebody has to design the simulation languages and their processors in the first place, we will take a brief look at the simulation process in this chapter by studying three simple examples, one continuous and two discrete.

A11.1 CONTINUOUS SIMULATION

Although continuous simulation finds most of its applications in engineering and science, many business and economic models use it as well. The central idea is that of values varying with time. For example, an economist might specify a model in which the increase in the gross national product is proportional to the total labor force and to the capital investment over the preceding year—that is, the rate of change in the gross national product is specified as a function of those other values.

In a typical engineering problem, space flight, the rate of change in the velocity of the spacecraft might be proportional to the power generated by the propulsion system. The objective of the simulation is to determine how the system behaves over time, and frequently the ultimate objective is to select values for particular variables to achieve a particular goal. The starship docking example in Chapter P3 is such a problem. Another simple example is examined below.

Example A11.1 *Rate of Chemical Reaction*

A chemist has two substances P and Q in one liter of water. They are combining to form a third substance R. The chemist knows that if the amounts of P and Q present in the water are p and q grams, respectively, then $2\alpha pq$ grams of R are formed each second. It is also known that each gram of R formed requires 0.5 grams each of P and Q. Thus if at time t seconds there are p, q, and r grams, respectively, of P, Q, and R in the water, then at time $t + 1$ seconds there will be $p - \alpha pq$, $q - \alpha pq$, and $r + 2\alpha pq$ grams, respectively. However, there is also a second effect occurring: in each second, $2\beta r$ grams of R are breaking up into βr grams each of P and Q. Thus the change in one second is really

- amount of P increases by $\beta r - \alpha pq$
- amount of Q increases by $\beta r - \alpha pq$
- amount of R increases by $2\alpha pq - 2\beta r$

Thus if $p(t)$, $q(t)$, and $r(t)$ are the amounts of P, Q, and R at time t, we have the following formulas for the amounts at time $t + 1$:

$$\begin{aligned} p(t+1) &= p(t) + \beta r(t) - \alpha p(t)q(t) \\ q(t+1) &= q(t) + \beta r(t) - \alpha p(t)q(t) \\ r(t+1) &= r(t) - 2\beta r(t) + 2\alpha p(t)q(t) \end{aligned}$$

This model can be simulated on the computer. A loop updates the amounts of p, q, and r once for each simulated second.

If we think about this model for a moment, we see that it is only an approximation to the truth: the chemical changes are actually occurring continuously, and as they occur the amounts of P, Q, and R change. Thus, although the amounts may be $p(t)$, $q(t)$, and $r(t)$ at time t, so that the *rate of production* of R is $2\alpha p(t)q(t) - 2\beta r(t)$ grams per second, the amounts will change before a second is over, so that the rate of production will also change.

Perhaps if we used half a second as our interval, we would obtain a more realistic model. Then we would have

$$p\left(t + \frac{1}{2}\right) = p(t) + \frac{\beta}{2}r(t) - \frac{\alpha}{2}p(t)q(t)$$

$$q\left(t + \frac{1}{2}\right) = q(t) + \frac{\beta}{2}r(t) - \frac{\alpha}{2}p(t)q(t)$$

$$r\left(t + \frac{1}{2}\right) = r(t) - \beta r(t) + \alpha p(t)q(t)$$

If this still is not an adequate model, we could reduce the interval in t still further, say to h seconds. Then we would have

$$\begin{aligned} p(t + h) &= p(t) + h\beta r(t) - h\alpha p(t)q(t) \\ q(t + h) &= q(t) + h\beta r(t) - h\alpha p(t)q(t) \\ r(t + h) &= r(t) - 2h\beta r(t) + 2h\alpha p(t)q(t) \end{aligned}$$

We hope that as h becomes smaller our model is more realistic, since we are assuming that rate of production to be constant for smaller and smaller intervals h. (We are also assuming that if z grams are being changed per second, then hz grams will be changed in h seconds.)

Fortunately it can be shown that as h gets smaller, the answers $p(t)$, $q(t)$, and $r(t)$ to such a model become more realistic. It can actually be shown that $p(t)$, $q(t)$, and $r(t)$ approach some continuous (smooth) functions $\hat{p}(t)$, $\hat{q}(t)$, and $\hat{r}(t)$ as h gets smaller, and that if $h\alpha$ and $h\beta$ are reasonably small, the numerical model described gives a very good approximation to these functions. We say that the numerical solutions *converge* to the continuous functions.*

Program A11.1 simulates the behavior of the chemical system by computing the values of p, q, and r every $1/m$ seconds for a total time of t_n seconds. This program is written for an interactive system: it outputs a request to the user whenever it needs more input. The user is given the option of specifying which variables should be changed from a previous simulation. This allows a person to sit at a terminal and experiment with the effects of various values for the parameters. The program is designed so that only values that change from the previous simulation

*In fact, they converge to the solution of the *differential equations*

$$\frac{dp}{dt} = \beta r - \alpha pq$$

$$\frac{dq}{dt} = \beta r - \alpha pq$$

$$\frac{dr}{dt} = -2\beta r + 2\alpha pq$$

The models described above can be used to solve such differential equations. Continuous simulation models usually give rise to differential equations that are solved by methods like these, but this subject is beyond the scope of a first course.

Program A11.1 ***Simulation of a simple chemical problem***

```
CHEMISTRY: program
    Concentrations of the substances P, Q, and R are computed. Initial
    values, reaction rates, and the time interval can be specified.
    real P=0.0,Q=0.0,R=0.0,DP,ALPHA=0.0,BETA=0.0
    integer M=1,TN=1,I,J,CONTROL=0
        do while CONTROL≠−1
        |   input CONTROL
        |   A control number of 0 will give the user instructions about
        |   what to do. Cases 0 through 4 are handled in a case state-
        |   ment. Other cases are ignored. A control number of −1
        |   will terminate the loop after this pass.
        |   case (0,1,2,3,4,) of CONTROL
        |   | 0: output 'CONTROL VALUES ARE'
        |   |    output '−1 STOP              0 PRINT THIS TABLE'
        |   |    output '1 INITIAL VALUES     2 NEW ALPHA, BETA'
        |   |    output '3 NEW M,TN           4 SIMULATE'
        |   | 1: output 'INPUT INITIAL VALUES OF P,Q,R'
        |   |    input P,Q,R
        |   | 2: output 'INPUT ALPHA,BETA'
        |   |    input ALPHA,BETA
        |   | 3: output 'INPUT # STEPS / SECOND,TOTAL TIME'
        |   |    input M,TN
        |   | 4: output 'ALPHA=',ALPHA,'BETA=',BETA
        |   |    output 'SIMULATION WITH',M,'STEPS PER
        |   |      SECOND'
        |   |    output 'TIME','P','Q','R'
        |   |    output 0,P,Q,R
        |   |    do for I←1 to TN
        |   |    |   Simulate for one second
        |   |    |   do for J←1 to M
        |   |    |   |   Compute rate of change in 1/M seconds
        |   |    |   |   DP←(BETA*R−ALPHA*P*Q) / FLOAT(M)
        |   |    |   |   P←P+DP
        |   |    |   |   Q←Q+DP
        |   |    |   |   R←R−2.0*DP
        |   |    |   |   enddo
        |   |    |   output I,P,Q,R
        |   |    |   enddo
        |   |   endcase
        |   enddo
    endprogram CHEMISTRY
```

need be specified. Consequently, initial values are specified for all variables that should be input by the user before the first simulation. This prevents the program from misbehaving if an input, such as t_n, is inadvertently left unspecified.

Program A11.1 *prompts* the user with a list of the items to be input. This method also provides a simple way of informing the user what control options may be specified. A *control number* indicates which variables the user wishes to change, but this arrangement is of no value unless the user can remember which number corresponds to which set of variables. It is a good idea in any interactive program to give the user a way to get a list of the available control options. Sometimes this can be done by testing for a special control input such as HELP. Since we do not want to use character strings in this program, we have chosen a control input of zero for this purpose. All the user has to remember is that an input of zero can be used to find out how to use the program.

A11.2 DISCRETE SIMULATION

Discrete simulation frequently deals with events of a random nature. For example, the arrival time of a customer at a bank is random *as far as the bank can observe*. Of course, the customer does not view the arrival as a random event, but as an event determined by other random variables, such as the amount of traffic. However, the best that the bank can do in planning its operations is to assume some sort of random arrival time. As discussed in Chapter A7, we will assume the existence of a random-number generator, RANDOM(), whose value is equally likely to be any floating-point value between 0.0 and 1.0. The random-number generator can be used to generate sequences of events for discrete simulation.

For example, suppose the bank decides that the model will have a customer arriving each minute with probability 1/10. This can be simulated by selecting a random number each simulated minute, and saying that a customer has arrived if the random number is less than 0.1 (because there is a 1/10 chance that the random number will lie between 0.0 and 0.1 rather than between 0.1 and 1.0). In simulation, a *simulated clock* keeps track of the simulated time: with this model, each time the simulated clock advances one minute, another random number is obtained to determine whether a customer has arrived.

Suppose we use this simple model of arrival times for customers at a gas station with one pump, and assume that it takes exactly eight minutes to serve each customer. How long will each customer have to wait on the average? Although the answer to very simple problems such as this one can be found mathematically, in more complex problems it is necessary to simulate numerically to find the answer. The numerical simulation of this problem will help us understand some of the techniques.

A program can calculate the average waiting time by computing the waiting time of each customer that arrives, summing those waiting times, and dividing by the number of customers. When the first customer arrives, there is no wait. However, the pump will now be busy for eight minutes, so any customer who arrives in the next seven minutes will have to wait until the pump is free. This suggests that we should keep a record of the service time remaining for all customers who have already arrived. In addition, we will need to know the total waiting time for all customers and the number of customers that have arrived. Program A11.2 stores these values in the variables SERVICE, WAIT, and CUST, respectively. Each simulated minute, the program reduces the

Program A11.2 ***Simple simulation model of a gas station***

```
GAS: program
    Compute the average waiting time of customers at a gas station
    which takes 8 minutes to service each customer. Customers arrive
    each minute with a probability of 0.1.
    real AVERAGE__WAIT, RANDOM
    integer N, WAIT, SERVICE, CUST, TIME
        input N
        SERVICE←0
        do while N>0
        |   WAIT←0
        |   CUST←0
        |   do for TIME←1 to N
        |   |   SERVICE←MAX(SERVICE−1, 0)
        |   |   if RANDOM()≥0.9
        |   |       then
        |   |           CUST←CUST+1
        |   |           WAIT←WAIT+SERVICE
        |   |           SERVICE←SERVICE+8
        |   |       endif
        |   |   enddo
        |   AVERAGE__WAIT←FLOAT(WAIT) / FLOAT(CUST)
        |   output 'AVERAGE WAIT FOR', N, 'CUSTOMERS IS',
        |     AVERAGE__WAIT
        |   input N
        |   enddo
    endprogram GAS
```

service time SERVICE by one (unless it is already zero) and determines whether another customer has arrived. If so, the wait for that customer is given by SERVICE, so it is added to WAIT, the number of customers is increased by one, and SERVICE is increased by eight. The simulation model is run for N simulated minutes, supplied as an input parameter.

The simulation can be repeated as many times as desired by providing additional values of N. It is terminated by giving a zero value for N. The reason for running the simulation more than once is to compare results. Since arrivals are random, each simulation may yield a different answer. If the simulation has been run for a suitably long time, the result will be close to the expected waiting time; but if it is run for only a short time, the result may be very different. When multiple runs of a simulation are made to compare answers in this way, it is important not to restart the program simply by resubmitting it to the computer. The random-number generator, RANDOM() in this program, does not actually produce random numbers: it produces numbers that are in a sequence that *appears* to be random. If the sequence is begun from the same place, the same results will be obtained. Most random-number generators have provisions for starting at different points in the sequence, but the details are machine- and language-dependent and will not be discussed here.

Program A11.2 uses a very simple model for the arrival times of customers. A more realistic model is based on *interarrival times*, the times between consecutive customer arrivals. The model might assume, for instance, that the time between two consecutive arrivals is a random number between 0.5 and 4.5 minutes, with any time being equally likely. (In practice, the assumption is that the interarrival time may be any time greater than zero, with shorter times being more probable than longer times—but this involves the theory of random distributions, a topic beyond the scope of this text.) If we want to construct a random number with values that are equally likely between any pair of values, we can use the random-number generator to generate a number between zero and one, and convert that range to the desired range. For example, we can generate a random number between 0.5 and 4.5 by computing 0.5+(4.5−0.5)*RANDOM().

In the next example we will use interarrival times for customers arriving at the drive-up windows of a bank. Suppose the bank managers are trying to decide how to lay out the approaches to their two drive-up windows. The alternatives are to provide two separate drives, one for each window, or a single drive feeding both windows. In the former case, a customer has to decide which line to join on arrival; in the latter case, the customer joins the single line, and, after arriving at the head of the line, takes the next available window. It is fairly obvious that the latter solution will lead to an average wait no longer than the former, and might lead to shorter waits, because there will be times when a customer joins one line only to find that the other line becomes empty sooner. However, if the latter solution is going to cost a little more, the bank managers may wish to determine the effect on the average wait to see whether the extra cost is justified. They may also want to measure some other statistics. For example, the customer's unhappiness increases more rapidly as the wait gets longer, so the managers may want to measure the average of the *square* of the waiting time. This will give an indication of the amount of variation in the waiting times: if the average of the squares is much larger than the square of the average waiting time, then there is a lot of variation in the waiting times for different customers.

In preparing a model for simulation, the bank managers must first determine the pattern of arrivals and service times for each customer. They do this by measuring the average number of customers arriving at various times of the day and the average length of time taken to service each customer. Let us suppose that these measurements have been taken, and know that the bank can expect customers to arrive randomly, with interarrival times between 2 and 6 minutes, and that service time is also random, between 1 and 10 minutes. To avoid complications, we will also assume that these random numbers are *uniformly distributed*—that is, that any value in the range is equally likely. With this much data, the bank managers can now ask, "What is the average waiting time and the average square of the waiting time, for the two arrangements?" The solution can be found by simulation.

Simulations of this type deal with *events*. An event in this case is the arrival of a customer or the completion of service for a customer. When an event occurs, the *state* of the system changes. When a customer arrives, the number of customers in line changes; when a teller completes the service of a customer, that customer leaves the system, and if there are customers waiting in line for that teller, a customer moves from the line for service. The basic action of the simulator is to keep track of the times at which events can occur, and advance the simulated clock to the next event time. For example, if the next customer is due to arrive in 2.6 minutes, but the next time that a teller will complete service is 3.5 minutes, the clock must be advanced forward 2.6 minutes and a new customer added to the line. A random number is then generated to determine when the next customer will arrive. If the time until the next arrival is 4.2 minutes, the next event will be the completion of service by a teller. The clock is advanced again and the next customer is removed from the waiting line. The service time for that customer determines the time of the next event concerning that teller. In this problem we must keep track of three event times: the time of the next customer arrival and the times at which each of the tellers will complete the service of a customer. The simplest way to do this is to store the time at

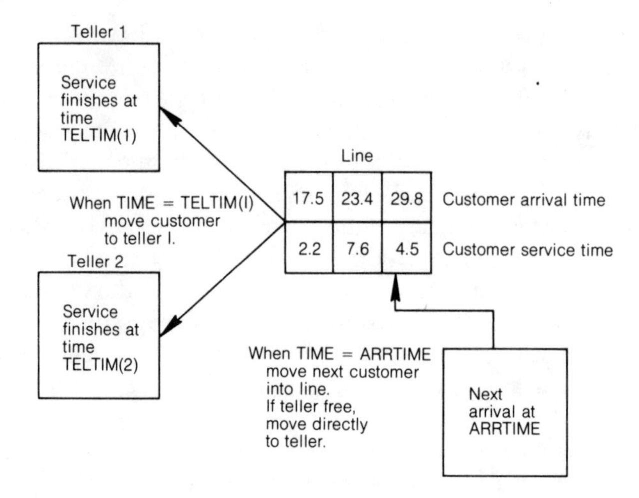

Figure A11.1a One-line system for two-teller bank

which each event will occur relative to the start of the simulation: that is, the simulated clock should be set to zero at the start of the simulation, and event times specified in minutes after that time. Figures A11.1a and A11.1b illustrate the two arrangements to be considered in this simulation. The assumption in the two-line system is that an arriving customer will choose the shorter line.

The basic structure of the simulation program consists of a loop that advances simulated time, held in the variable TIME, to the next event time and processes this event. The loop must be terminated by some mechanism. In this case, as in most simulations, we want to stop after

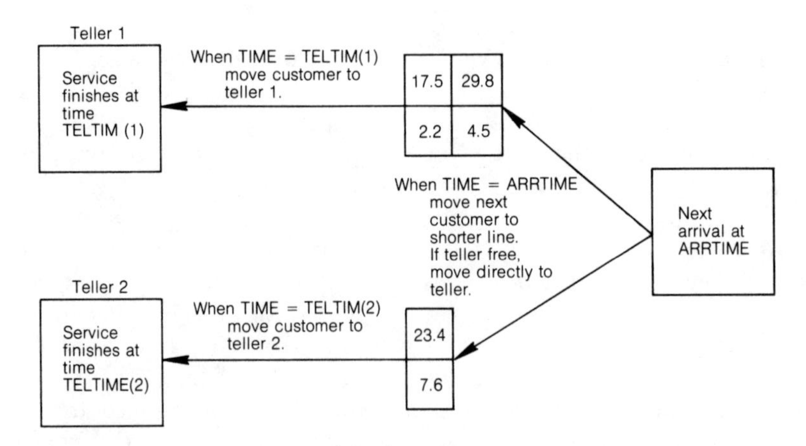

Figure A11.1b Two-line system for two-teller bank

some specified elapsed simulation time, which we hope will be long enough to give meaningful statistics. An outline of the program is shown in Program A11.3. This program inputs a parameter specifying the length of the simulated run. Subsequent inputs cause further simulations starting from the state left by the previous run—that is, there may be customers in line and being served. Unless the first run is very long, the initial start-up may bias the results, because it may take a while for the lines to build up to their normal length. Therefore, it is a good idea to run the simulation for a while to allow the lines to build up, and then to ignore the results from that first run.

Program A11.3 does not address the problem of how to store information about customers in the line or lines, nor does it consider how to

Program A11.3 ***Outline of bank simulation***

```
BANK: program
    A two-teller bank is simulated to compute expected wait times and
    squared wait times.
    real TELTIM(2),TIME,ARRTIME,STOPTIME,LENGTH,WAIT,WAITSQ
    integer NCUSTS
        initialize tellers and lines to empty
        input LENGTH
        TIME←0.0
        do while LENGTH>0.0
          | STOPTIME←TIME+LENGTH
          | WAIT←0.0
          | WAITSQ←0.0
          | NCUSTS←0
          | do while TIME<STOPTIME
          |   | find earliest event from ARRTIME and TELTIM(I),
          |   |   I=1,2
          |   | if next event is an arrival
          |   |   | then
          |   |   |     TIME←ARRTIME
          |   |   |     handle new arrival
          |   |   | else
          |   |   |     TIME←TELTIM(I)
          |   |   |     handle completion for teller I
          |   |   | endif
          |   | enddo
          | output WAIT / FLOAT(NCUSTS),WAITSQ / FLOAT(NCUSTS)
          | input LENGTH
          | enddo
    endprogram BANK
```

```
handle completion for teller I: subprogram
    (declarations)
        if customer waiting for this teller
            then
                move customer to teller from line
                TELTIM(I)←TIME+service time for this customer
                WAIT←WAIT+TIME−customer arrival time
                WAITSQ←WAITSQ+(TIME−customer arrival time)↑2
                NCUSTS←NCUSTS+1
                mark teller I as busy
            else
                mark teller I as not busy
            endif
        return
    endsubprogram handle completion for teller I
handle new arrival: subprogram
    (declarations)
        compute time of next arrival and assign to ARRTIME
        compute service time
        add to line (or to shorter line if two-line system)
        if teller I free for I=1 or 2
            then handle completion for teller I
            endif
        return
    endsubprogram handle new arrival
```

indicate that a teller is not busy. The latter problem can be handled in two ways: we can use either a flag to indicate whether each teller is free or a value of TELTIM that would not otherwise be used. Using special values of variables frequently leads to shorter programs, but not necessarily to programs that are easy to follow. Therefore, we will use the first method: we introduce the logical array BUSY, which is **true** in position I if teller I is busy and **false** otherwise.

The problem of storing information about customers requires more thought. As customers enter the system, we must keep a record of their arrival times and the time it will take to service them. The number of customers waiting will vary during the simulation, and in the two-line system there can be more customers in one line than in the other. This is a common situation in simulation: as *entities* (in this example, customers) move through the system, they must be saved in lines waiting to be processed by the next stage. In this example there is only one stage, the teller; but in a model of an automobile assembly plant, for example, there may be many stages, representing the different processes that must be completed in the manufacture of a car. The best mechanism for keeping track of the customers in a line is the *chained list* introduced in

Chapter A4. Each customer becomes one entry in a list representing the line. One list head is needed for each line, but if there are two lines, both lists can be stored in one set of arrays. The lists in this problem must contain two data items, customer arrival time and service time, so three arrays are needed—CUSTAR, SERVTM, and PTR—for the two times and a pointer. PTR(1) and PTR(2) can be used as the list heads for the lines for tellers 1 and 2 in the two-line system.

Program A11.4 gives a complete program for the simulation of the two-line system. The changes needed to simulate a one-line version are small: the first executable statement in the subprogram TELLER should be changed from K←I to K←1, and the statement containing if QLENGTH(2) . . . should be removed from the subprogram ARRIVAL.

Program A11.4 ***Simulation of a bank***

```
BANK: program
    A two-teller bank is simulated to compute expected waiting times.
    Variables used include
    TELTIM(I)              Time at which teller I will next be free
    ARRTIME                Time at which next customer will arrive
    BUSY(I)                Teller I is busy if true
    QLENGTH(I)             Number of customers waiting for teller I
                             including one at window
    PTR,CUSTAR,SERVTM      Storage for list structure containing point-
                             ers, arrival times, and service times of
                             customers.
    PTR(I), I = 1, 2       List heads for lines 1 and 2
    LEND(I)                Ends of lists for lines 1 and 2
    FREE                   Start of free list in list structure
        real TELTIM(2),TIME,ARRTIME,STOPTIME,LENGTH,WAIT,
          WAITSQ,CUSTAR(100),SERVTM(100),T
        integer QLENGTH(2),LEND(2),PTR(100),FREE,I,NCUSTS
        logical BUSY(2)
        global BUSY,TELTIM,ARRTIME,QLENGTH,PTR,CUSTAR,
          SERVTM,LEND,FREE,TIME,WAIT,WAITSQ,NCUSTS
            call INITIALIZE
            input LENGTH
            TIME←0.0
            do while LENGTH>0.0
              | STOPTIME←TIME+LENGTH
              | WAIT←0.0
              | WAITSQ←0.0
              | NCUSTS←0
              | do while TIME<STOPTIME
```

```
        T←ARRTIME
        I←0
        if T>TELTIM(1) and BUSY(1)
            then
                I←1
                T←TELTIM(1)
            endif
        if T>TELTIM(2) and BUSY(2) then I←2 endif
        if I=0
            then
                TIME←ARRTIME
                call ARRIVAL
            else
                TIME←TELTIM(I)
                call TELLER(I)
            endif
        enddo
    output 'NUMBER OF CUSTOMERS,AVERAGE WAIT,
      AND AVERAGE SQUARE WAIT'
    output NCUSTS, WAIT / FLOAT(NCUSTS),WAITSQ /
      FLOAT(NCUSTS)
    input LENGTH
    enddo
  endprogram BANK
TELLER: subprogram (I)
    Called if Teller I has just completed a customer, or is free and a
    customer has just been added to line K.
    (all global declarations from main program)
    integer I,J,K
        K←I
        if BUSY(I) then QLENGTH(K)←QLENGTH(K)−1 endif
        if PTR(K)≠0
            then
                J←PTR(K)
                PTR(K)←PTR(J)
                TELTIM(I)←TIME+SERVTM(J)
                WAIT←WAIT+TIME−CUSTAR(J)
                WAITSQ←WAITSQ+(TIME−CUSTAR(J))↑2
                NCUSTS←NCUSTS+1
                BUSY(I)←true
                if PTR(K)=0 then LEND(K)←K endif
                Return J to free storage.
                PTR(J)←FREE
                FREE←J
```

```
            else
                BUSY(I)←false
            endif
        return
    endsubprogram TELLER
ARRIVAL subprogram
    Handle new arrival. Compute service times and time of next arrival
    and add new customer to line. If no free space available, force
    TIME past stop time.
    (all global declarations from main program)
    integer I,J
    real RANDOM
        if FREE=0
            then
                output 'NO FREE SPACE LEFT, SIMULATION
                  ABANDONED'
                TIME←1.0E50
            else
                I←FREE
                FREE←PTR(I)
                ARRTIME←TIME+2.0+4.0*RANDOM()
                SERVTM(I)←1.0+9.0*RANDOM()
                J←1
                if QLENGTH(2)<QLENGTH(1) then J←2 endif
                Add to line J.
                QLENGTH(J)←QLENGTH(J)+1
                PTR(I)←0
                CUSTAR(I)←TIME
                PTR(LEND(J))←I
                LEND(J)←I
                if not BUSY(1)
                    then call TELLER(1)
                    else if not BUSY(2) then call TELLER(2) endif
                    endif
            endif
        return
    endsubprogram ARRIVAL
INITIALIZE: subprogram
    (all global declarations from main program)
    integer I
        do for I←1 to 2
            LEND(I)←I
            PTR(I)←0
            QLENGTH(I)←0
```

```
        BUSY(I)←false
        enddo
    do for I←3 to 99
        PTR(I)←I+1
        enddo
    PTR(100)←0
    FREE←3
    ARRTIME←0.0
    return
endsubprogram INITIALIZE
```

Problems

1. An automatic speed control for an automobile has the following characteristics:

   The velocity of the vehicle V is compared with the desired velocity VDES. The difference DIFF = VDES−V is used to adjust the accelerator position AP every second. The adjustment consists of an increase in AP by an amount equal to C*DIFF−D*AP where C and D are coefficients to be determined by simulation. The automobile is assumed to obey a mathematical model in which its speed can be calculated each second from the equation

   VNEW = V+AP−F*V

   that is, if the speed at time T seconds is V and the accelerator position is AP, the speed at time T + 1 seconds is as given. The new accelerator position is given by

   APNEW = AP+C*(VDES−V)

   Write a program to simulate the behavior of the automobile, assuming that the frictional force coefficient F is 0.5 and the desired velocity VDES is an input parameter. Try various values of the coefficient C between 0.1 and 0.8 to see what happens to the control system.
2. Some simulations require a mixture of continuous and discrete techniques. Consider a simulation of the temperature of a building controlled by an automatic heating system and subject to random comings and goings of people. A model of such a system has a continuous part, representing the temperature of the building and the rate of flow of heat into and out of it, and a discrete part, representing the turning on and off of the thermostat and the random arrivals and departures of people. Suppose we have the following model:

The temperature of the building at time T + 1 is given by the temperature of the building at time T, the heat input H from the heating system, the heat loss LOSS through the walls, and the heat E generated by other equipment.

$$\text{TEMPERATURE}_{new} = \text{TEMPERATURE}+\text{H}-\text{LOSS}+\text{E}$$

The heat input H is either zero, if the heating system is off, or 0.63, if the system is on. The system is shut off at time T + 1 if the temperature at time T is greater than TEMPDESIRED + 1; it is turned on at time T + 1 if the temperature at time T is less than TEMPDESIRED − 1.

LOSS is given by 0.03*(TEMPERATURE−OUTSIDETEMP)

E is either 0 or 0.11, and changes from one value to the other as a person enters or leaves. For the model, E can be assumed to change from one value to the other with probability 0.1 each unit of time—that is, if at time T, E is 0.11, then at time T + 1, E should be 0.11 with probability 0.9 and 0.0 with probability 0.1.

Write a program to simulate this system assuming that TEMPDESIRED = 24 and OUTSIDETEMP = −5. Find the *duty cycle* of the heating system—that is, the percentage of the time that it is on.

***3.** Program A11.4 uses lists to store the line(s) of waiting customers. These lists are manipulated in subprograms TELLER and ARRIVAL: the former removes an element from a list and returns the space to the free list, the latter gets space from the free list and adds an element to a list of customers. In general, it is not good programming practice to manipulate lists in subprograms whose principal task is something else. In this case, TELLER is principally concerned with simulating the action of the teller, and ARRIVAL with the arrival process. Therefore, Program A11.4 should be organized differently: separate subprograms should be written to add an element to the end of a list and to remove an element from the front of a list. These subprograms can then be called from ARRIVAL and TELLER, respectively. (This organization also has the advantage that, when additional aspects of the bank's operation need to be simulated, the list-manipulation subprograms are available for use by other modules.)

Rewrite the subprograms TELLER and ARRIVAL following this superior organization. (Notice that ARRIVAL has to take care of the difficulty that arises when there is no space in the free list. Since the simulation cannot continue in this case, the subprogram prints an error message and then forces termination by advancing TIME to a value larger than any value that will be used. This is a decision that should be made in the subprogram ARRIVAL, not in a list-manipulation subprogram, since there could be other procedures calling the list-manipulation subprogram that want to take some action other than terminating the process when there is no free space left.)

Chapter

A12

Trees, Queues, and Stacks

In Chapter A4 we encountered a data structure called a *chained list*, or *list* for short. The advantage of a list is that elements can be inserted in it or removed from it at any point without having to move large amounts of data in memory. There are many applications in which we want to represent relationships between data items—such as ordering (item A is less than item B), family relationships (entry A is the parent of entry B), and connection (point A is connected to point B), just to name a few examples. An ordering relationship can be described by a list, but other relationships involve more complex information—for example, one parent can have several children, or one point can be connected to many other points.

Often it is necessary to represent the *structure* of the data in the computer, so that the structural relationships can be used in solving the problem. Even when the initial data has very little structure, it may be desirable to provide additional structure during the course of the solution in order to implement a fast method. This chapter introduces three very important types of *data structure*: the *tree*, the *queue*, and the *stack*. These structures are inherent in many problems, and are used in the solution of even more. They will be introduced by simple examples: later chapters in this module present important applications of these structures to real problems.

A12.1 BINARY TREES

Lists were introduced as a useful mechanism for maintaining sets of unordered or ordered items that are to be updated by insertions and deletions. If a list is unordered, the only way to search for an item is

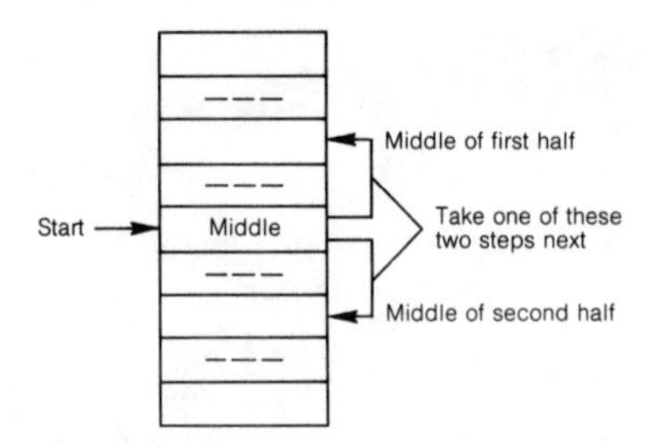

Figure A12.1 Binary search

sequentially; but we saw in Chapter A3 that an ordered list allows us to use the fastest method of searching large sets of data, the *binary search.* Unfortunately, Program A3.2 cannot be used on chained list, because the program refers to the Ith entry in the list. Finding the Ith entry in a chained list requires a number of operations proportional to I. In a binary search, we go to the middle entry in the table. If this is not the entry sought, we go to the middle of the first or second half of the table, depending on whether the middle entry is larger or smaller than the one sought. This process is diagrammed in Figure A12.1.

The figure suggests that we could use a pointer to tell us where the middle of the table is located, and then use two more pointers to tell us where the middle parts of the two halves of the table are located. If, after going to the middle of one half of the table, we find that the entry there is not the desired one, we wish to continue to the middle of a half of the half. Again, we could find that entry if we had two more pointers. This idea is illustrated in Figure A12.2 for a table with seven entries in alphabetical order. Null pointers have been omitted for clarity. Each entry now consists of three items: the data and two pointers, one to the earlier part of the table and one to the later part. The structure becomes clearer if drawn as in Figure A12.3. Each entry is shown as a triplet LEFT, DATA, and RIGHT. LEFT is a pointer to the part of the table to the left of the entry, RIGHT a pointer to the part of the table to the right of the entry. This type of data structure is called a *tree.* If it is drawn "up the other

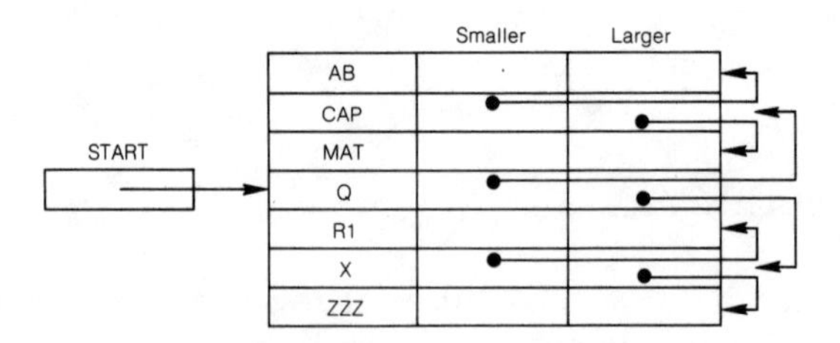

Figure A12.2 Search with pointers

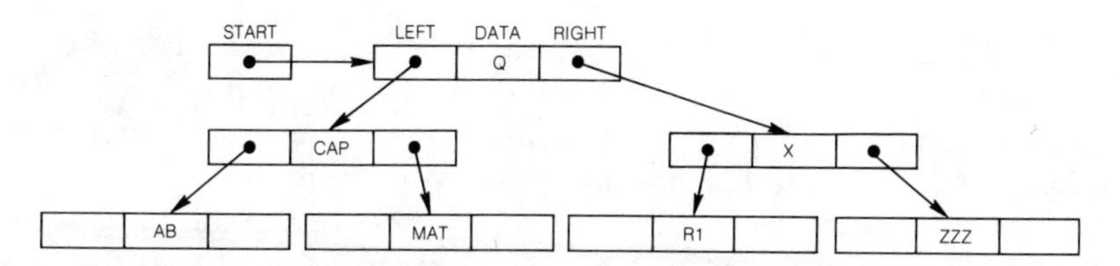

Figure A12.3 Expansion of Figure A12.2

way," as shown in Figure A12.4, the analogy is obvious, but we will normally draw trees as in Figure A12.3.

Much of the terminology of tree structures is based on horticulture. As shown in Figure A12.4, the bottom (top in Figure A12.3) of a tree is called the *root*, the places where we have entries are called *nodes*, the pointers from one node to other nodes are called *branches*, and the nodes with no pointers leaving them are called *terminal nodes* or *leaves*. Formally, a tree is a set of nodes connected by branches in such a way that there is one and only one way of going from one node to another via branch connections, and which has a distinguished node called the *root node*.

Another part of tree terminology is derived from genealogical trees. Referring to Figure A12.3, the nodes below and directly connected to a given node are called the *offspring* of that node (also called *subnodes*). Thus CAP and X are offspring, or subnodes, of Q. If we start with CAP in Figure A12.3 and consider the part of the tree connected to CAP and not above it, we have another tree, called a *subtree* of the original tree. In this case, the subtree consists of CAP, AB, and MAT. The nodes in the subtree are all *successors*, or *descendants*, of Q. If a node A has a node above it and directly connected to it, say node B, then B is the *parent* of A. The only node without a parent is the root node. If two nodes have the same parent, they are said to be *siblings*. For example, AB and MAT in Figure A12.3 are siblings.

The tree in Figure A12.3 is a special type of tree called a *binary tree*, in which each node has exactly two offspring, referred to as the *left* and

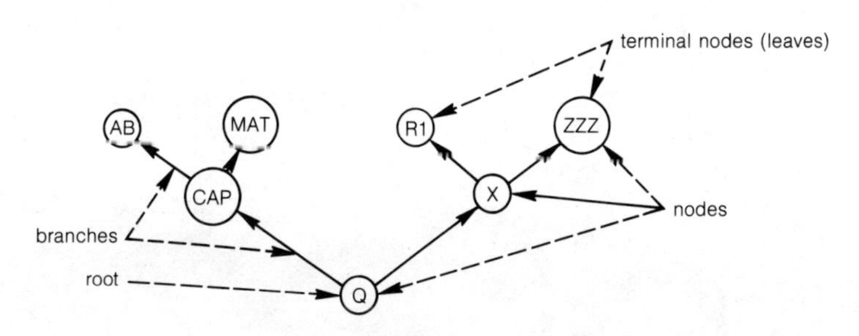

Figure A12.4 Tree

right offspring. Either or both offspring of a given node may be null. A binary tree is not simply a tree that has no more than two subnodes for any given node. For example, a family tree showing maternal relationships in which no mother has more than two children is not a binary tree, because, in the case of a mother with only one child, there is no distinction between "left" and "right." In binary trees, this distinction is important. If, for example, the entry ZZZ is removed from Figure A12.3, node X has a null right offspring. The tree is *ordered,* and this property would be destroyed if the subnode R1 were moved from the left to the right of X.

Binary trees are particularly important in computer applications because they are easy to represent using three arrays. The important property of the tree in Figure A12.3 is that all nodes connected via the left pointer of a given node are alphabetically less than that node. Similarly, all nodes connected via the right pointer are alphabetically greater. If we wish to search for a given item, say R1, we start at the root and compare R1 with the entry there. Since R1 is larger, it is either not in the tree or connected to the right pointer. Therefore we proceed through the right pointer of the root. There we find that R1 is less than the entry X. We therefore proceed via the left pointer and arrive at R1.

Example A12.1 *Searching an Ordered Binary Tree*

Searching an ordered binary tree can be programmed very simply. The code is shown in Program A12.1. This code is much simpler than that in Program A3.2, because there is no need to construct addresses for the

Program A12.1 *Search ordered binary tree*

```
BINARY__TREE__SEARCH: subprogram (I,DATA,LEFT,RIGHT,ROOT,
  X,N)
    This subroutine searches the tree rooted in ROOT. The triplets
    representing the tree are stored in LEFT(J), DATA(J), RIGHT(J). If a
    J is found such that DATA(J) = X, I is set to J; otherwise I is set to 0.
    integer I,N,DATA(N),LEFT(N),RIGHT(N),ROOT,X
        I←ROOT
        do while I≠0 and DATA(I)≠X
            if DATA(I)>X
                then I←LEFT(I)
                else I←RIGHT(I)
                endif
            enddo
        return
    endsubprogram BINARY__TREE__SEARCH
```

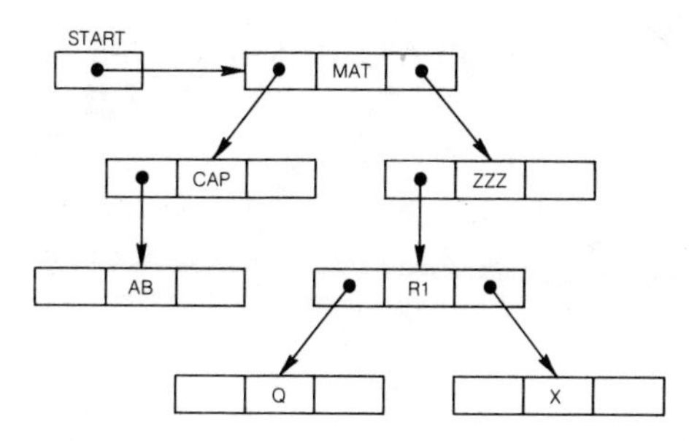

Figure A12.5 Alternative tree for storing data in Figure A12.3

array entries: they are given directly in the pointer arrays LEFT and RIGHT. The program assumes that DATA is an array of integers, but it can be used for any data type by changing the declarations for DATA and X. The variable ROOT contains the index of the root of the tree. Although the table in the example above is "balanced," it is not necessary that the root node contain exactly the middle entry or that any node contain the middle entry of the section of the tree below it (that is, the *subtrees* connected to it via its left and right pointers). But all entries accessible via the left pointer of any node must be alphabetically less than the entry in the node, and all entries accessible via the right pointer must be alphabetically greater: that is, the tree must be *ordered.* If this relationship is maintained, then the search algorithm given in Program A12.1 will work.

As an example, the data given in Figure A12.3 could be stored using the tree structure shown in Figure A12.5. The tree is still ordered, so that when we search for R1, we can compare it with MAT in the root node and determine that it must lie in the subtree to the right of that node if it is in the table. The difference between Figures A12.3 and A12.5 is only one of average execution time. Suppose each of the seven entries were to be looked up. The number of levels of the tree that must be searched is shown in Table A12.1. We can see that a balanced tree leads to fewer operations on the average if any item is equally likely to be referenced. The extreme case occurs when all nodes have a null subtree on one side, as shown in Figure A12.6. The number of levels of search necessary for this case is also given in Table A12.1. The tree in Figure A12.6 is equivalent to a list, since there is only one branch from each node. Consequently, a search in this tree is equivalent to a sequential search: we expect the average time to be about N/2, where N is the number of entries.

Because a binary tree does not have to remain balanced, it is easy to add a new entry. Suppose we have the configuration shown in Figure A12.5 and we wish to add the string CAR. We search down the table looking for CAR until we reach a null pointer. If we were to find CAR, we

TABLE A12.1 LEVELS OF SEARCH NECESSARY

Entry	Levels of Search		
	Balanced Tree in Figure A12.3	Unbalanced Tree in Figure A12.5	Worst-Case Tree in Figure A12.6
AB	3	3	1
CAP	2	2	3
MAT	3	1	4
Q	1	4	7
R1	3	3	6
X	2	4	5
ZZZ	3	2	2
Total	17	19	28
Average	2.43	2.71	4.00

couldn't add it, because our structure does not allow for repeated entries. In this example we find a null pointer when we attempt to go right from CAP, so we add CAR to the tree at that point. This occurs as a result of the following sequence of steps:

1. Compare CAR with MAT. CAR is less, so go left.
2. Compare CAR with CAP. CAR is greater, so go right.
3. Right contains a null pointer, so stop and put CAR on the right of CAP.

Successive stages starting from an empty table and adding the strings CAN, AC, MAT, Z1, and I3, are shown in Figure A12.7.

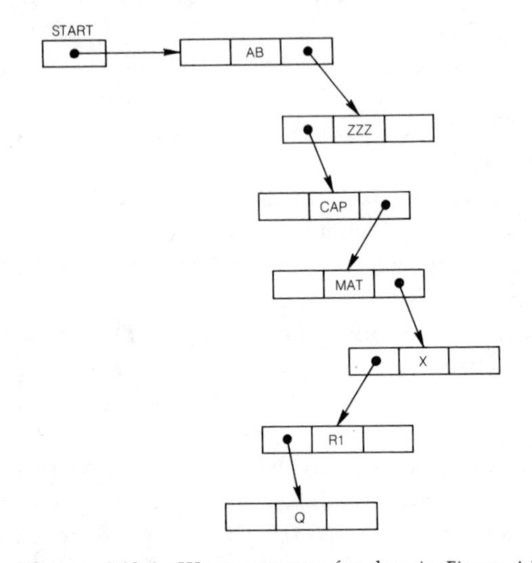

Figure A12.6 Worst-case tree for data in Figure A12.3

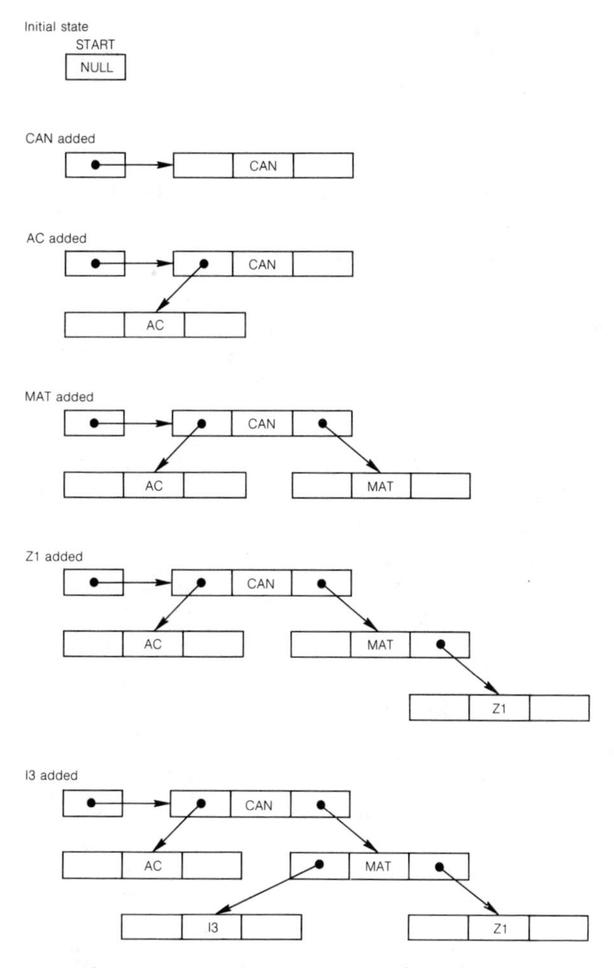

Figure A12.7 Successive stages of tree building

As new items are added to a tree, storage space for the triplets can be obtained from a storage-allocation mechanism using a free list (in this case a list of triplets). If items are to be deleted from a tree, the space freed can then be returned to the free list.

Deletion from an ordered tree is more difficult. Unless the deleted entry is a terminal node, some reorganization of pointers is necessary. If either the right or the left pointer of the deleted entry is null, the problem can be handled as in a list (see Figure A12.8). In all other cases we must replace the deleted entry with an entry from either its left or its right subtree. Suppose we decide to use an entry from the right subtree. It must be the smallest entry in that subtree, as shown in Figure A12.9.

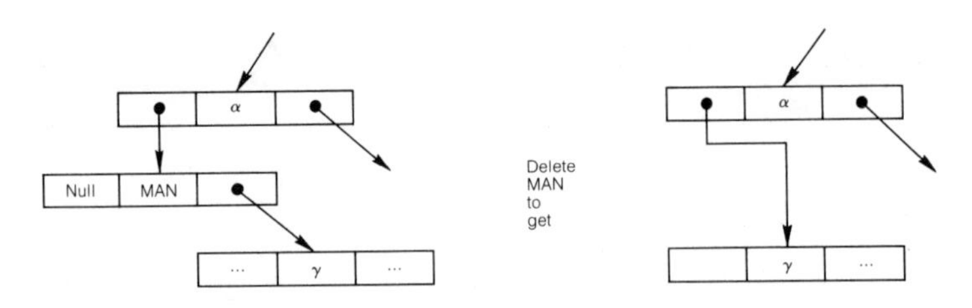

Figure A12.8 Deletion from an ordered tree

Program A12.2 performs the operation shown in Figure A12.9. This example is worth studying, because it illustrates the importance of selecting the proper representation for the data structure. When we try to program deletion, we find that special cases complicate the program and make it difficult to follow. Suppose we started by trying to write a subprogram to remove the node with index I. It is clear that the contents of the parent node of I must be modified: if we do not give the index of the parent as a parameter, the entire tree must be searched to find the parent. (If we are given the *data* to be removed, a search for the node containing that data will also find the parent.) If the item to be removed is the root node, there is an immediately obvious special case: there is no parent. This special case can be eliminated by storing the index of the root in one of the pointer variables, say LEFT(1), making it appear that the root has a parent at index 1.

Because the node to be removed may be either the left or the right offspring of the parent, it is necessary to indicate which. Program A12.2 requires that this be done by using a negative pointer value if the node to be removed is the left subnode, a positive pointer value otherwise.

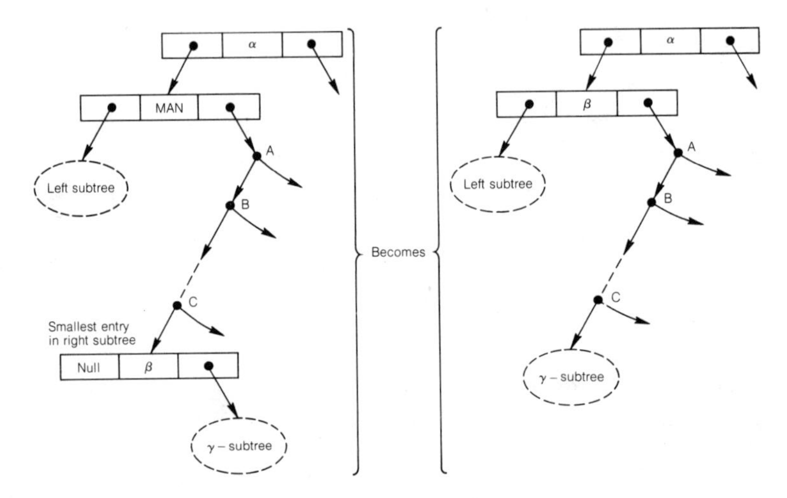

Figure A12.9 Deletion when replacement necessary

Program A12.2 ***Delete a node from a binary tree***

```
DELETE__NODE: subprogram (I,LEFT,RIGHT,N)
   The node that is the offspring of the node at index ABS(I) is removed
   from the binary tree with pointers stored in arrays LEFT and RIGHT.
   If I is negative, the left offspring is to be removed, otherwise the right.
   integer I,N,LEFT(N),RIGHT(N),J,K,L
      J is set to the index of the node to be removed, L to the index
      of the node to be put in its place—that is, I will point to L instead
      of J at completion.
      if I<0 then J←LEFT(−I) else J←RIGHT(I) endif
      if LEFT(J)=0 or RIGHT(J)=0
         then
            If either subtree of J is null, remove J immediately as
            in Figure A12.8.
            L←LEFT(J)+RIGHT(J)
         else
            Go down right subtree of J to find leftmost node L
            (node β in Figure A12.9). Set K to point to parent of
            node L with same sign convention as used for I.
            K←J
            L←RIGHT(J)
            do while LEFT(L)≠0
               K←−L
               L←LEFT(L)
               enddo
            Move L to position occupied by J by replacing pointers
            in L. Make K point to offspring of L.
            if K<0
               then LEFT(−K)←RIGHT(L)
               else RIGHT(K)←RIGHT(L)
               endif
            RIGHT(L)←RIGHT(J)
            LEFT(L)←LEFT(J)
         endif
      Finally, make I point to L.
      if I<0 then LEFT(−I)←L else RIGHT(I)←L endif
      return
   endsubprogram DELETE__NODE
```

A12.2 STACKS AND QUEUES

Chained lists allow elements to be inserted or deleted at any position. Two very important special classes of lists are *stacks* and *queues*. A *queue* is a list with the restriction that entries may be added or deleted at either end (but not in the middle); a *stack* is a queue with the further restriction that items may be added or deleted at one end only. These structures are especially useful, first because they arise naturally in many applications, and second because they can be implemented very efficiently. A queue was used in the bank simulation example in Section A11.2. New arrivals at the bank were placed at one end of a list representing the line of waiting customers, and removed from the other end when it was their turn to be served. This type of queue, in which entries are added at one end and removed at the other, is called a *first-in-first-out* or *FIFO* queue. Stacks have the property that the entries are removed in the *reverse* of the order in which they are entered. Thus, if the entries X, R, and T are placed in a stack, they are removed in the order T, R, and X. For this reason, a stack is called a *last-in-first-out* or *LIFO* queue. In this section we will consider an application of stacks and then show how stacks and queues can be implemented efficiently.

Stacks are particularly useful in processing data structures such as trees, so we will examine a common problem concerning trees. We often wish to examine the entries of a tree in a particular order. For example, suppose we want to print the contents of an ordered binary tree in alphabetical order.

Let us consider the tree in Figure A12.10: a, b, and c are the addresses of locations where the tree elements are stored (or are their indices in an array used for storage). We start at the node containing B, but before we can print that, we must print all entries in the subtree to the left of B. Therefore we move left to A. Since that has no subtree to its left, we can print A. Now, since there is no subtree to the right of A, we must move back to B and print it. But how can we get back to B? We have no pointer to show us the way back. When we moved down to A from B, we should have saved the address b of B.

Now consider the tree in Figure A12.11. We move left to A2 from B, saving its address b. There is a subtree to the left of A2, so we must print that subtree before printing A2. Hence we move left from A2, saving its address a2. As we go deeper into the tree, we will accumulate more and more addresses that must be saved so that we can find our way back. How should we save these addresses? A stack structure is exactly what we want, because when we return up the tree, we will want to

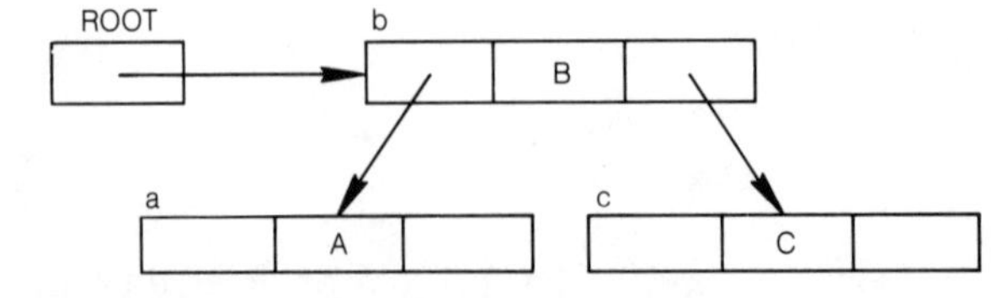

Figure A12.10 Tree with three entries

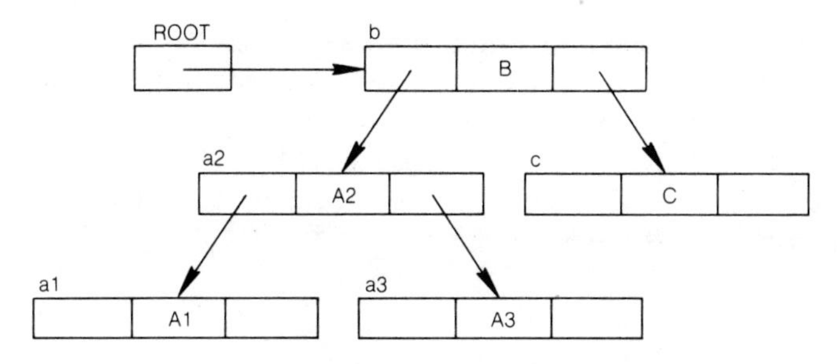

Figure A12.11 Tree with five entries

revisit all nodes in reverse order. Each time we go deeper into the tree, we put the address of the entry we are leaving on top of the stack. When, for example, we have reached A1 in Figure A12.11, the stack will be as shown in Figure A12.12. When we wish to return to the next higher level, A2, its address is on top of the stack. After we remove it, the address of the level above, B, is on top of the stack.

Suppose we have just printed A1. We return to the next higher level by removing the address a2 from the top of the stack. Now we can print A2. Next we wish to go to the right subtree of A2 and print it. What should we save in the stack in order to return? After we have printed the right subtree, we do not wish to return to A2, because we have already printed it. Instead, we wish to go straight to B and print it. Thus when we go down to the right of A2 we need not put a return into the stack. The address b already there is the appropriate place to which to return.

The two important operations on a stack are *push* and *pop*. *Push* puts another entry on top of the stack, *pop* removes the entry currently on top of the stack, making it one entry smaller. Program A12.3 prints an ordered binary tree rooted in ROOT in increasing order. It uses the push and pop operations, written as push X onto stack and pop stack into X. The latter functions like an assignment statement, changing the value of X; the former simply copies the value of X onto the top of the stack. The binary tree is assumed to be stored in LEFT, DATA, and RIGHT as before.

Conceptually, we can visualize a stack as if new storage cells were added to the top of the stack during a push operation and removed during a pop operation. A chained list can be used to implement this

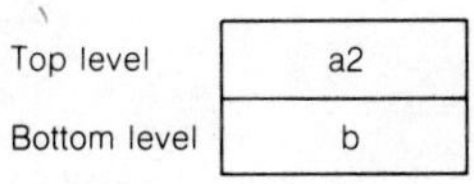

Top level a2
Bottom level b

Figure A12.12 Stack with two entries

Program A12.3 ***Print an ordered binary tree***

```
PRINT__TREE: subprogram (ROOT,LEFT,DATA,RIGHT,N)
    integer ROOT,N,LEFT(N),DATA(N),RIGHT(N),PTR
    The tree rooted at ROOT is printed in order. A stack keeps track of
    the path into the tree.
        PTR←ROOT
        set stack empty
        do while PTR≠0 or stack not empty
            GO__LEFT: do while PTR≠0
                This loop goes down to the left until a NULL pointer
                is found.
                push PTR onto stack
                PTR←LEFT(PTR)
                enddo GO__LEFT
            pop stack to PTR
            output DATA(PTR)
            PTR←RIGHT(PTR)
            enddo
        return
    endsubprogram PRINT__TREE
```

visualization. The "new storage cell" for a push operation can be obtained from a free list; after a pop operation a cell must be returned to the free list. However, a stack can be implemented more efficiently using an array. Suppose we know that no more than 150 entries will be present in the stack at any one time. We can dimension an array STACK to have 150 entries, and assign STACK(1) to hold the bottom entry, STACK(2) to hold the next entry, and so on. The location of the top of the stack must be known, so another integer variable, STACK__TOP, is needed to hold the index of the top of the stack, as shown in Figure A12.13. The operation push PTR onto stack used in Program A12.3 can be coded as

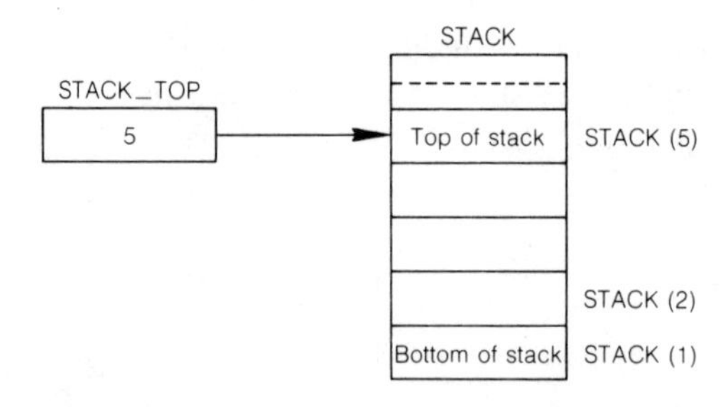

Figure A12.13 Stack implemented in an array

```
if STACK__TOP<150
    then
        STACK__TOP←STACK__TOP+1
        STACK(STACK__TOP)←PTR
    else
        output 'STACK OVERFLOW IN PUSH OPERATION'
    endif
```

The pop stack to PTR operation can be coded as

```
if STACK__TOP>0
    then
        PTR←STACK(STACK__TOP)
        STACK__TOP←STACK__TOP−1
    else
        output 'STACK UNDERFLOW IN POP OPERATION'
    endif
```

Arrays have several advantages over chained lists for implementing stacks. Less memory space is used, because pointers do not have to be stored. Furthermore, about half as many operations are needed in the push and pop operations when arrays are used. The main disadvantage of arrays is that their size must be fixed before the stack is first used, so it is necessary to know the maximum size of the stack ahead of time.

Arrays can also be used to implement queues. Because data can be added or deleted from either end, the lower end of the queue cannot be assumed to be in a fixed place, as in a stack. Consequently, the indices of both ends of the queue must be kept. Figure A12.14a shows an array QUEUE being used to hold a queue whose two ends are in QUEUE(FIRST)

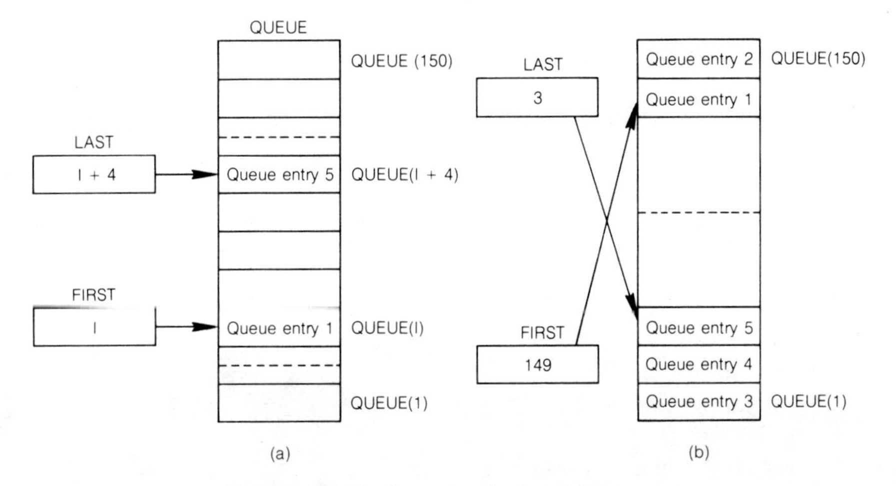

Figure A12.14 Queue implemented in an array

and QUEUE(LAST), respectively. If a number of entries are added to the FIRST end, FIRST will be decremented, and may eventually go below 1. In this case we must store entries in the top end of the array, as shown in Figure A12.14b, where the queue contains 5 entries. We can also arrive at the state shown in Figure A12.14b by adding a large number of entries to the LAST end while removing them from the FIRST end: when the LAST end of the queue reaches the top of the array, it must "wrap around" to the lower end of the array.

*A12.3 RECURSION AND STACKS

In Chapter P11 it was pointed out that stacks are a natural mechanism for allocating memory space in recursive procedures—so, not surprisingly, problems that use stacks can often be solved recursively in a natural way.

Recursion, for example, can greatly simplify Program A12.3. An ordered binary tree is printed by first printing the left subtree of the root in order, then printing the value in the root node itself, then printing the right subtree in order. This amounts to a recursive definition, provided that the "simple" case is handled nonrecursively. In this example, the simple case is a null tree, which requires no printing at all. Program A12.3a is a recursive version of A12.3 based on this definition. The advantage of the recursive version is its simple structure, which makes it easy to write and test.

Program A12.3a ***Print an ordered binary tree***

```
PRINT__TREE: subprogram (ROOT,LEFT,DATA,RIGHT,N)
    The binary tree rooted in ROOT is printed recursively in ascending
    order.
    integer ROOT,N,LEFT(N),DATA(N),RIGHT(N)
        If ROOT is null, do nothing but return; otherwise print the left
        subtree, followed by the data at the root, and the right subtree.
        if ROOT≠0
            then
                call PRINT__TREE(LEFT(ROOT),LEFT,DATA,RIGHT,N)
                output DATA(ROOT)
                call PRINT__TREE(RIGHT(ROOT),LEFT,DATA,RIGHT,N)
            endif
        return
    endsubprogram PRINT__TREE
```

*Chapter P11 should be read before this section.

Problems

***1.** What is the maximum number of entries that can be stored in a binary tree if the longest path from the root to any node does not exceed n? What is the average search time for any entry in that tree when n is 3, 4, or 5? Can you give the average for any n?

2. If Program A12.3a is used to print the tree shown in Figure A12.3, what is the sequence of calls on PRINT_TREE? (Give the value of the parameter ROOT for each call.)

3. Write a recursive program to search for an entry X in an ordered binary tree.

***4.** Suppose you have to write a program to input a set of words and their definitions. (Each word and its definition can be viewed as a character string.) The program must also accept words without definitions as input, in which case it is to print the definition of the word if one was previously input. Finally, the program is to output a list of all words input in alphabetical order with definitions. Would you use a binary tree to order the words alphabetically, or one of the sorting methods discussed in Chapter A3? Consider two different cases: (1) all the words with definitions are input before any of the words without definitions; (2) the definitions and the words to be looked up can appear in the input in any order.

5. Write two subprograms which implement a FIFO queue using an array QUEUE as shown in Figure A12.14. Remember that items can only be added to the LAST end and removed from the FIRST end. Your codes should check for queue overflow or underflow (trying to add another entry when the array is full, or to remove an entry when the array is empty). Remember to allow for FIRST to contain a larger index than LAST, as shown in Figure A12.14b.

Chapter

A13

Polish Notation

In this chapter we are going to introduce another way of writing expressions that has the advantage that parentheses are never needed. In conventional expressions we use a hierarchy of operations in which, for example, multiplication takes precedence over addition. If we wish to write an expression in which addition is to be done before multiplication, we use parentheses, as in (A+B)*C. The purpose of the parentheses is to override the natural hierarchy of the operators. In *Polish notation* all operators have the same precedence: their position indicates the order of evaluation. This will be seen to have many benefits for the computer. People are used to writing expressions in the form used so far in this book; we are not suggesting that they should change. However, it is convenient for the computer to translate the usual form of expression into Polish form during the process of compiling a program.

Let us first consider trees as a way of representing the order in which the operations in an expression are to be executed. Note that each operation, such as multiplication, uses two operands. We have previously called these *binary* or *dyadic* operations. For the moment we will forget about the *unary* or *monadic* operations such as −A. We can represent each binary operation by a tree with three nodes and two branches, as shown in Figure A13.1. The root node contains the operator and its

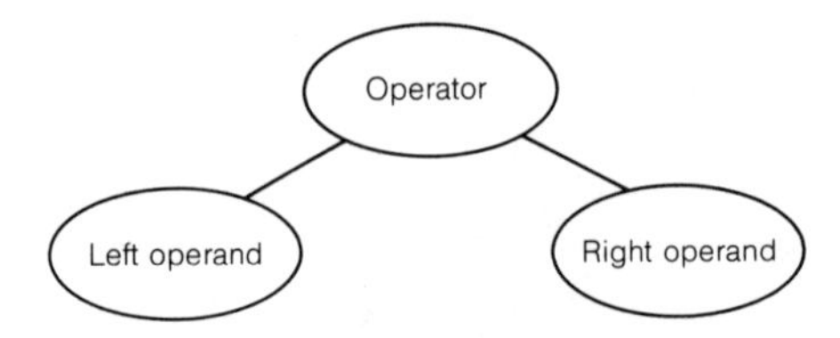

Figure A13.1 Tree representing an arithmetic operation

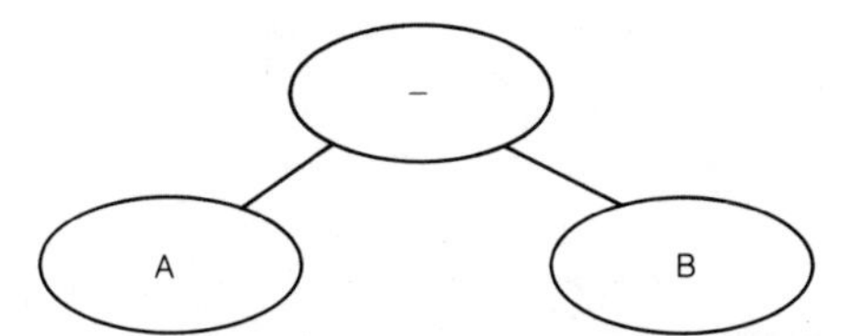

Figure A13.2 Tree for A−B

two subnodes contain the two operands. Thus A−B would be represented as shown in Figure A13.2.

When we come to a more complex expression such as P−Q*R, we note that the hierarchy of operators requires that the multiplication use the two operands Q and R. Thus we can construct a three-node tree for that operation. The subtraction operator uses P as its left operand and the result of the multiplication as its right operand. Thus we can construct a tree for this operation if we can find something to use for the right operand. If we think of a tree as standing for the value of the expression it represents we can use the tree representing the multiplication operation as the right subnode of the subtraction tree. Thus P−Q*R is represented as shown in Figure A13.3. In this tree the order of the operators is determined by the fact that the multiplication must be performed first in order to provide the right operand for the subtraction: it is not dependent on any hierarchy assigned to the operators themselves. Thus if we wish to draw a tree for the expression (P−Q)*R, we can start by drawing the tree for P−Q and then use it as the left operand for the multiplication operation, as shown in Figure A13.4. The fact that the multiplication operator has a higher precedence than the subtraction operator is irrelevant, since the structure of the tree itself indicates that the subtraction must occur first. Consequently we do not need parentheses when we represent expressions as trees.

Notice that the tree representing an expression has operators in its

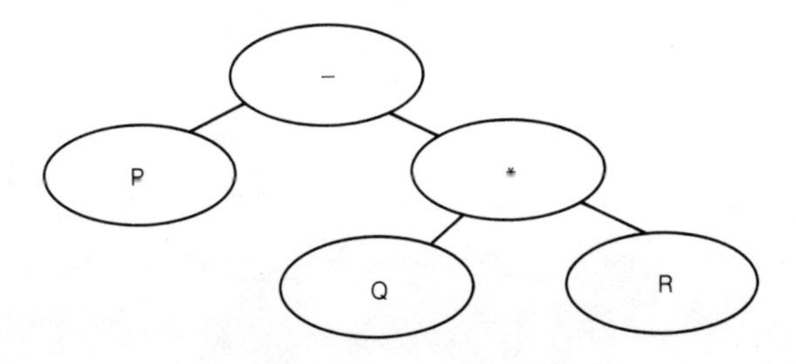

Figure A13.3 Tree for P−Q*R

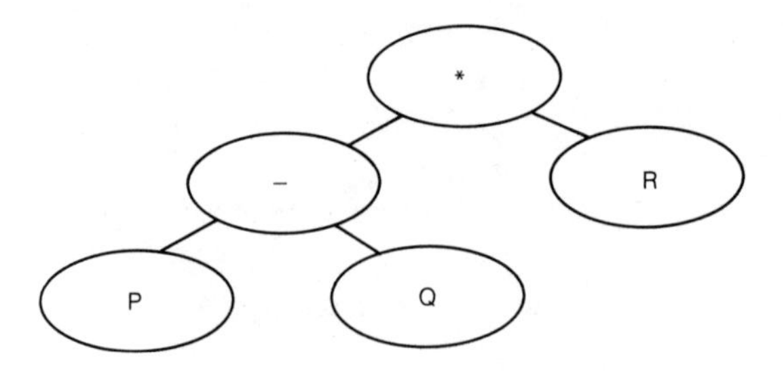

Figure A13.4 Tree for (P−Q)*R

nonterminal nodes and operands in its terminal nodes. This will always be true, since each node (except the root node) is an operand for the node above, so it either consists of data, such as P, or is a subtree that will calculate the operand. Tree structures for some more complex expressions are shown in Figure A13.5. Notice that each subtree is itself an expression, whose value is an operand for the parent node of the root of that subtree.

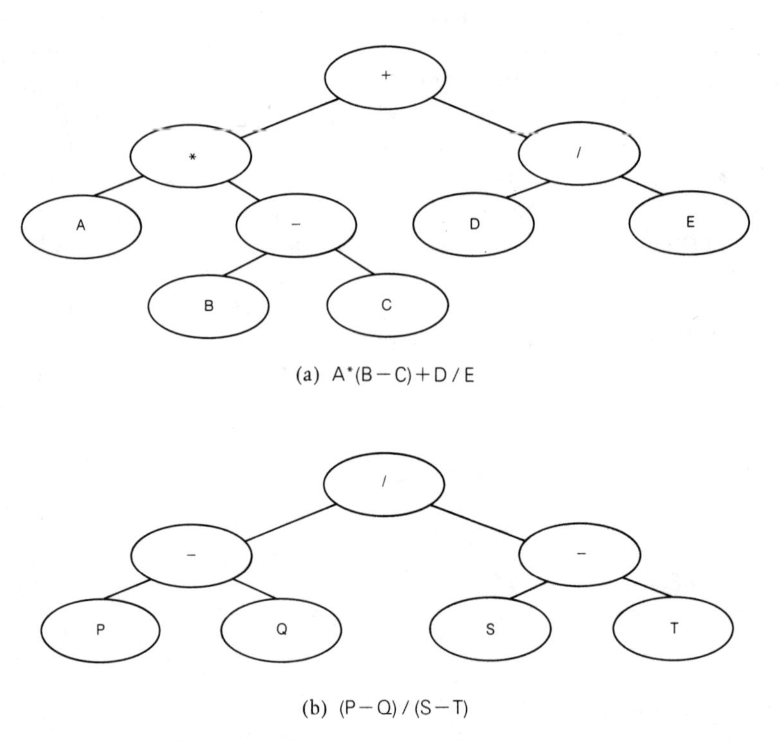

(a) A*(B−C)+D/E

(b) (P−Q)/(S−T)

Figure A13.5 Expressions and their trees

We can use trees to represent expressions without using parentheses and without requiring a hierarchy of operators. However, a tree is not a convenient way of storing information concisely, because, as we saw in Chapter A12, we need to store many pointers along with the information. If, however, we can find a way of writing trees in a compact form, we can use them to write expressions without parentheses. The Polish form is such a method.

Recall, from Chapter A12, that when we wanted to print a tree in order we visited each node by starting at the root node and following down to the left until we got to a terminal node. We then printed that and came back up to the node above and printed that, followed by the information to the right of it. If we applied this process to the tree in Figure A13.5a, we would print the string A*B−C+D/E. This is the original expression with the parentheses removed. Hence the order of the operations has been lost. However, if we print the tree in a different order, we will find that we can reconstruct the order of the operations. In this new order we visit each node as before, starting at the root node, but we print the information in the itself node first, then the information to the left of it, followed by the information to the right of it. Thus we would print the tree in Figure A13.1 as "operator," "left operand," "right operand." The tree in Figure A13.2 would be printed as −A B. This is interpreted as "the operator subtract is to operate on the operands A and B by subtracting the second from the first." Whereas we are used to writing A−B for this operation, in the new form we write the operator first. This is called *Polish prefix* notation, because it was invented by the Polish mathematician Jan Lukasiewicz; "prefix" refers to the fact that the operator precedes the expression. (Our regular notation is called *infix* notation, because the operator is between the two operands.) If we write the tree in Figure A13.5a in prefix form, we get

+*A−B C/D E

Although this looks very strange to us, it is possible to convert it back into a unique tree and hence understand what it means.

We convert a prefix expression back into a tree by realizing that each operator has two operands. Hence if we start with the string

+*A−BC/DE
123456789

(the entries have been numbered so we can refer to them in the text below) we look at the first entry (+) and then start looking for its two operands. At this time we start a tree with the plus operator as the root node. Its left subnode will be its first operand, so we look for that first. The second entry in the string is a multiplication operator. Hence it will occupy the left subnode of the root, as shown in Figure A13.6a. Since it

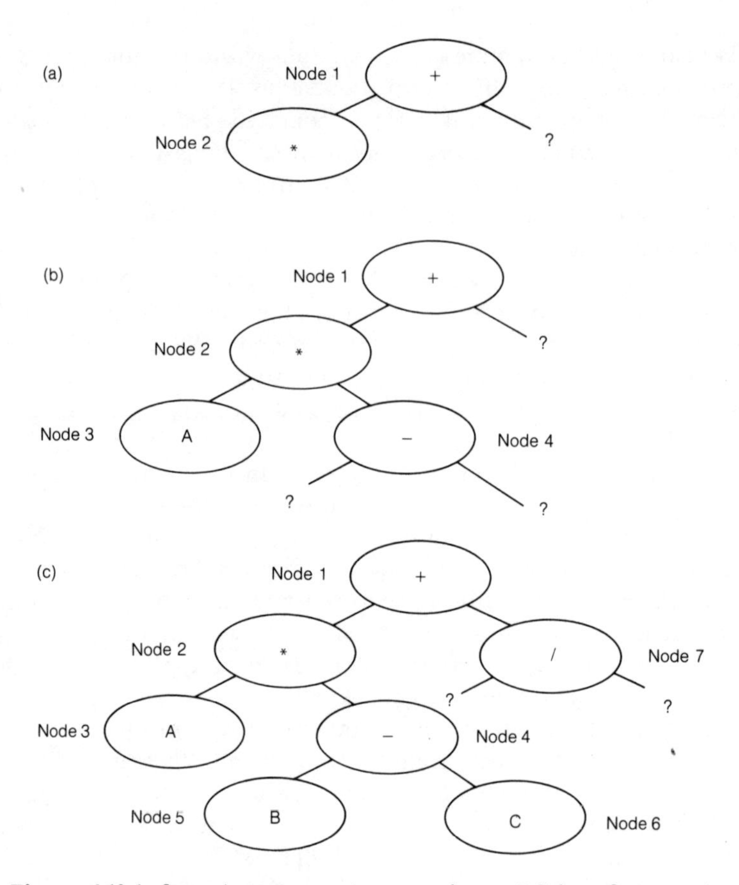

Figure A13.6 Stages in reconstructing a tree from a Polish prefix expression

is an operator, we temporarily abandon our search for the right subnode of the root and begin looking for two operands for the multiplication in node 2. The third element in the string is A, so this will be the left operand for node 2.* Now we can look for the right operand for node 2. The fourth entry in the string is another operator, subtraction. This becomes the right subnode of node 2, and we now start looking for two operands for it. The state of the tree at this point is shown in Figure A13.6b. The fifth and sixth entries in the string are operands, so these are the left and right subnodes of node 4. Now we have finished constructing a left operand for node 1, and we can start looking for a right

*We assume that each character in the Polish string represents an element (that is, either an operator or a piece of data). Many compilers have a "prescan" section that groups the input characters into elements. We will use single characters for these elements unless they are numeric data, in which case the group of characters forming one element will be clearly delineated.

operand. The seventh entry in the string is another operator, so it is the right subnode of node 1. We must look for a left and right subnode for it. We now have the tree shown in Figure A13.6c. The next two entries in the string are operands, so we finally recover the tree shown in Figure A13.5a.

Another way of understanding the meaning of a prefix expression is to draw an underline from each operator to its two operands. We can start this process by working from the end of the expression and moving back to the last operator. A line can be drawn from it under the following two operands. In the example we have been using we get

$$+ * A - B\ C\ \underline{/\ D\ E}$$

This means that the division operator operates on D and E. Scanning back to the left, we next come to the subtraction. A line can be drawn from it to the operands B and C to get

$$+ * A\ \underline{-\ B\ C}\ \underline{/\ D\ E}$$

Scanning back to the left still further, we come to the multiplication operator. This time its first operand is A and the next is the group $\underline{-\ B\ C}$. Thus we get

$$+\ \underline{* A\ \underline{-\ B\ C}}\ \underline{/\ D\ E}$$

Finally we scan back to the left to reach the initial addition. Its two operands are the two groups $\underline{* A\ \underline{-\ B\ C}}$ and $\underline{/\ D\ E}$. In this process we recognize a group as a single operand because we have underlined it.

This process required that we read the Polish string from the end to the beginning, but often this is inconvenient. Instead we can use *Polish postfix*, in which the operator is written *after* its two operands. Thus the infix expression A−B would be written A B− in postfix notation. This corresponds to reading through the tree in the same order as before, but writing the left and right operands before the operator. Using the same example (Figure 13.5a), we start at the root node containing the addition operator. First we must write the left operand, so we follow down to the left and come to the multiplication. Before we can write it, we must write its two operands. The left is A, so it can be written immediately. The right is a subtree starting at the subtraction, so its operands must be written first: they are B and C. Now we come back to the subtraction node: since both its operands are now written, we can write down the subtraction operator itself. So far we have the string A B C−. We back up to the multiplication node and can now write it down. This brings us back to the root node containing the addition, but we still have to write down its right operand. This results in the three entries D E /; then we can finally write down the addition to get:

A B C−*D E/ +

If we wish to reconstruct the tree from postfix form, it is easier to start from the right-hand end, where the root appears. However, we can read from left to right to find out which are the operands for the various operators. We read to the first operator and draw a line back to the previous two operands. This results in

A B C−*D E/ +

telling us that the subtraction operates on B and C. Now we draw a line back from the multiplication. Its operands are A and the group already underlined. This gives

A B C−*D E/ +

The scan continues and we come to the division. Its operands are D and E, so we underline those to get

A B C−*D E/ +

Finally we come to the plus sign, and find that its operands are the two groups A B C−* and D E/.

We have not mentioned unary operators, such as the minus sign in D*(−D+E). In fact, there are three different uses of the symbol − in an expression. In A−B*(−C+ −3.5), the first − symbol is the familiar binary subtraction operator. The second is the unary minus sign, meaning that we are to use the negative of the value of C. The third is part of the number-naming process: that is, the string −3.5 names a certain number that is to be used as an operand in the addition operation. This use is not as an operator at all. In normal use we write the same symbol for all three cases.

We cannot use one symbol for several purposes in Polish notation. If we see the symbol − as an entry in a string, for example, we will look for its two operands. Therefore, we must use a different symbol for the unary minus. Let us use the lowercase letter m. The tree corresponding to A−B*(−C+ −3.5) is shown in Figure A13.7. The binary subtraction in node 1 has two subnodes, whereas the unary minus in node 6 has only one subnode. The prefix and postfix forms for this tree are

−A*B+m C −3.5

and

A B C m −3.5+*−

respectively. Notice that the group −3.5 is a single entry in the Polish string.

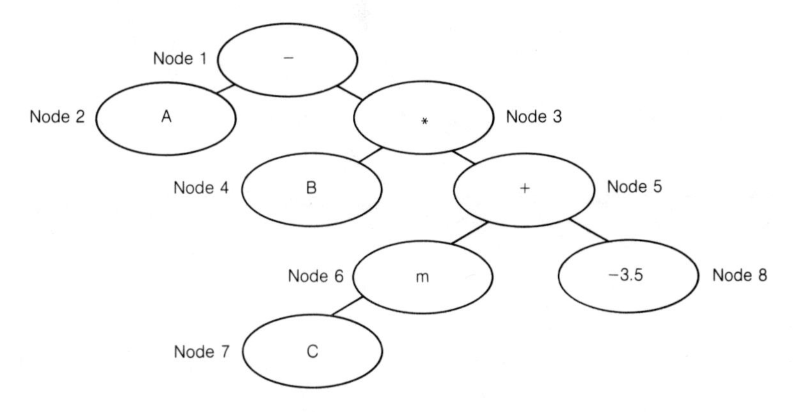

Figure A13.7 Tree for A−B*(−C+−3.5)

As before, we can convert the forms back into their tree representations or we can understand them by underlining the groups. We get

+A*B+m C −3.5

and

A B C m −3.5+*−

respectively, for the prefix and postfix forms of the example above.

When Polish strings are stored inside a computer, they will be encoded in such a way that each entry is separated from the next. For example, operators will probably be represented by integers, possibly 1 for +, 2 for −, 3 for *, and so forth; data will probably be represented by the addresses of table entries describing that data (for example, the name of the variable, whether it is an integer or a real, etc.).

A13.1 EVALUATING POLISH EXPRESSIONS

An important advantage of Polish postfix expressions is that it is a simple matter for a computer program to evaluate them. In this section we will study the automatic evaluation of expressions, represented first as trees and then as Polish postfix expressions. In the following section we will examine a method for converting the usual infix expressions into Polish postfix form.

Suppose we wish to find the value of the tree shown in Figure A13.8a. It is evident that we must start by executing an operator that has terminal nodes for both operands. One way we can find such an operator is to

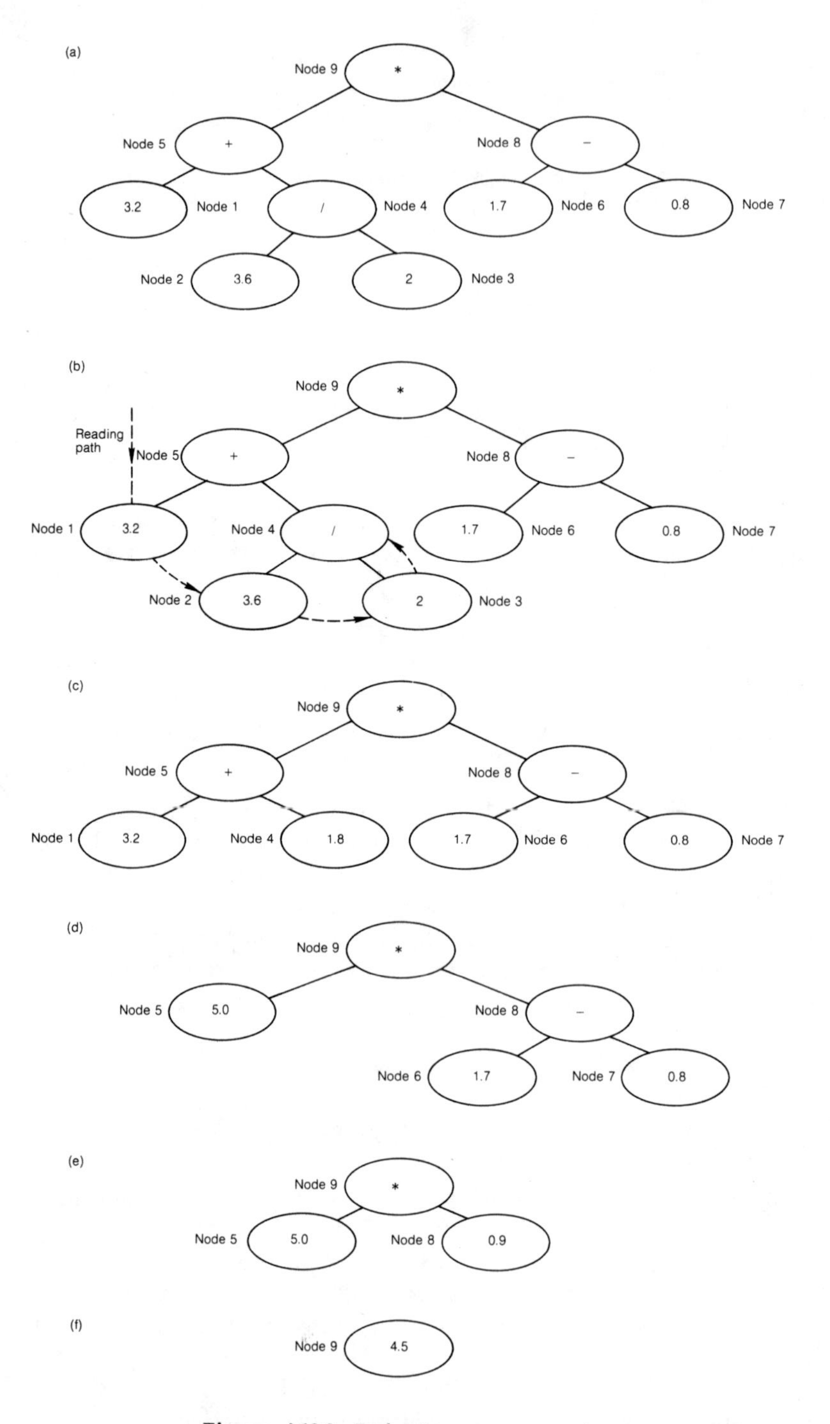

Figure A13.8 Evaluating a tree expression

read the tree in the usual left-to-right fashion, starting from the root. We should read each node containing an operator in the order "left subnode, right subnode, operation." If a subnode in turn contains an operator, it must be read in the same fashion before the reading of its parent can continue. Eventually we will come to an element with both subnodes terminal—that is, containing data. In Figure A13.8a this is the division in node 4, which will be read in the order "left subnode (3.6), right subnode (2), divide." The reading path is shown by the dashed arrows in Figure A13.8b. At the time we first read an operator, the last two items read were its operands. Therefore we can apply the division operator to the operands 3.6 and 2 to get the result 1.8. The subtree rooted on the division operator can then be replaced by a terminal node containing 1.8. We have removed one operator and simplified the tree.

An identical process can now be applied to this simplified tree. Again we will find that the first time we complete the reading of a node in the form "left subnode, right subnode, operator," we will just have read the two operands for the operator. The successive stages of such an evaluation are shown in Figures A13.8c to A13.8f.

The mechanism just proposed for evaluating a tree expression requires that the expression be scanned repeatedly from left to right to look for an operator "at the bottom" of the tree. The rescanning is not necessary if we keep track of the path we have already followed in reading the tree. Look back at Figure A13.8c and notice the path that has been followed in reading the tree. Now notice that after the division has been performed, the two nodes read prior to the division have been removed, and the result of the division (1.8) has replaced the division operator in its node. If we draw what remains of the reading path of the tree in Figure A13.8c, we get the path shown in Figure A13.9a. If we continue reading we will come back up to the addition operator, as shown in Figure A13.9b. This path is exactly the one that would be followed if we started reading the reduced tree from the beginning. Now the preceding two nodes on the reading path are the operands for the addition. If we perform this operation, the tree and the reading path are collapsed as shown in Figure A13.9c.

The reading path consists of a list of nodes that have already been read. It will contain only data in all places except possibly the last. Even there, the presence of an operator is temporary, because it is immediately executed, using the two preceding entries on the list as operands. Since we work only on the end of the list, it is in fact being used as a *stack* (see Chapter A12.) Instead of building a list of nodes that have been read, we could put their contents into a stack. In the example of Figure A13.8b, the state of the stack after having read the first three nodes on the reading path would be as in Figure A13.10a. When the next node

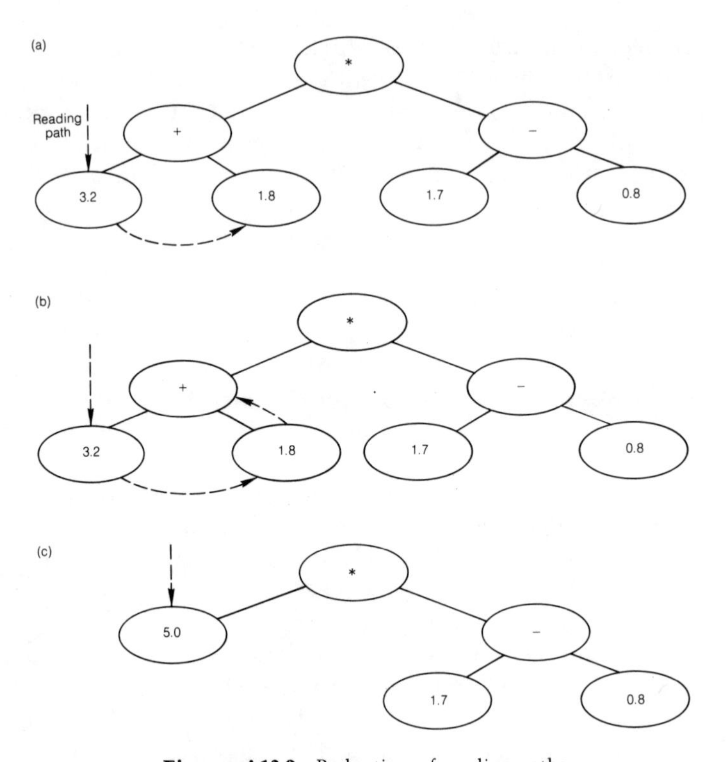

Figure A13.9 Reduction of reading paths

(containing the division) is read, the last two entries on top of the stack are combined by dividing the lower by the upper. The result is put back in the stack, as shown in Figure A13.10b. The next node to be read contains the addition operation. It is immediately executed on the top two entries of the stack to get the result shown in Figure A13.10c. Successive states of the stack before and after reading each of the remaining operators are shown in the remainder of Figure A13.10. The final state of the stack is a single entry containing the value of the original expression.

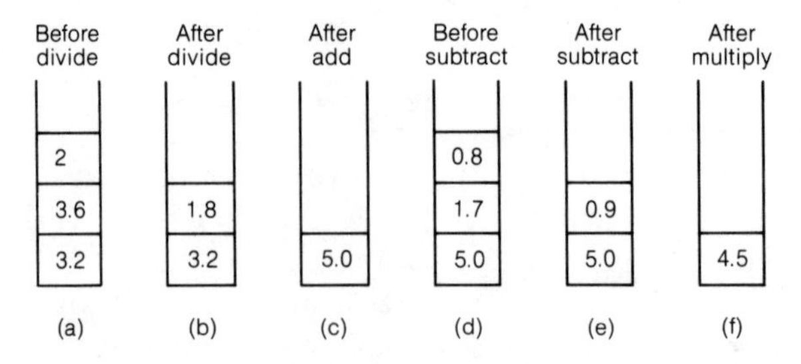

Figure A13.10 States of a stack

Expression remaining to be read	Stack (top →)		
31.5,2,3.5,*,m,/,3,1.5,/,+			
2,3.5,*,m,/,3,1.5,/,+	31.5		
3.5,*,m,/,3,1.5,/,+	31.5	2	
*,m,/,3,1.5,/,+	31.5	2	3.5
m,/,3,1.5,/,+	31.5	7	
/,3,1.5,/,+	31.5	−7	
3,1.5,/,+	−4.5		
1.5,/,+	−4.5	3	
/,+	−4.5	3	1.5
+	−4.5	2	
	−2.5		

Figure A13.11 Evaluation of a postfix string

At this point we note that the order in which we have been reading the tree is precisely that of a Polish postfix expression: the two operands are read, followed by the operator. Thus if we are given such an expression, we now know how to evaluate it using a stack. Whenever an operand is read, it is placed in the stack. When an operator is read it is executed, using the top entries in the stack as data and returning the result to the top of the stack. The evaluation of 31.5 / −(2*3.5)+ 3 / 1.5, represented by the postfix expression 31.5,2,3.5,*,m, /,3,1.5, /,+, is shown in Figure A13.11. The operation m is unary minus; the items in the expression have been separated by commas for clarity. Notice that the unary minus is executed between the fifth and sixth lines of Figure A13.11. Since it is a monadic operator requiring only one operand, it uses only the top item in the stack and returns the result to the same place.

A13.2 SYNTAX ANALYSIS AND CONVERSION TO POSTFIX

The *syntax* of a language is the structure of that language, and *syntax analysis* is the determination of the structure of a statement in that language. Thus, if the language is English, a syntax analysis of the statement

The boy hits the ball.

is used to determine that boy and ball are nouns, the first being the subject of the sentence and the second the object. It then can be used to see that

<subject> <verb> <object>

is a valid sentence. If the language is a programming language, a syntax analysis is used to determine the structure of a string such as A←B+C. We first determine that B and C are operands, then see that B+C is an expression that is a valid right-hand side, and that A is a valid left-hand side, and finally determine that

<left-hand side>←<right-hand side>

is a valid assignment statement.

In this section we are going to study the problem of converting an infix expression into its equivalent postfix form. To do this we have to determine the syntax of the expression. If there are syntax errors, as in the expression B* / C, we cannot find a meaningful structure. In this case a compiler analyzing the expression should not come to an abrupt termination, but should localize the error as much as possible before continuing to scan the input for other errors.

The basic operation in converting infix to postfix is that of moving the operator from its position between the two operands to a position following them. Thus A+B becomes A B+. Hence if we are presented with a string A+B and we read it from left to right, we must save the addition operator when we read it until after we have read its second operand. However, we can write out the operands immediately as we read them. Now let us consider the infix string

A+B↑C*D / E

How do we determine that the operands for the multiplication are B↑C and D? We must use the hierarchy of the operators shown in Table P2.1.

Because exponentiation has a higher precedence than multiplication, the variable C, which is immediately to the left of the multiplication, is an operand for the exponentiation rather than for the multiplication. B↑C is an operand in turn for either the addition on its left or the multiplication on its right. Since addition has a lower precedence than multiplication, B↑C is an operand for the multiplication. Similarly we look to the right of the multiplication for its second operand. Division has the same precedence as multiplication. Since our rule for operators having equal precedence is that the left operator is evaluated first, D is an operand for the multiplication. Using this approach, we can convert the expression by reading it from left to right. The process will be described below and is illustrated in Table A13.1.

First we read the variable A and output it into the postfix string. Then we read the addition operator. The preceding operand must be the first operand for the addition. However, we must save the addition until we have found its second operand and have output this operand into the postfix string. We continue reading and output the next operand B into

TABLE A13.1 CONVERSION OF INFIX TO POSTFIX

Input String (Infix)	Output String (Postfix)	Stack (Top→)	Precedence Comparison
A+B↑C*D/E△			
+B↑C*D/E△	A		
B↑C*D/E△	A	+	
↑C*D/E△	A B	+	
C*D/E△	A B	+↑	↑>+
*D/E△	A B C	+↑	
*D/E△	A B C↑	+	*<↑
D/E△	A B C↑	+*	*>+
/E△	A B C↑D	+*	
/E△	A B C↑D*	+	/=*
E△	A B C↑D*	+/	/>+
△	A B C↑D*E	+/	
△	A B C↑D*E/	+	△</
△	A B C↑D*E/+		△<+

the postfix string. Next we read the exponentiation operator. Since exponentiation has a higher precedence than the saved addition, the last variable output (B) is the first operand for the exponentiation. Therefore we must also save the exponentiation until we have found its second operand. Next we read C and output it to the postfix string. Then we read the multiplication operator, which has a lower precedence than the exponentiation that was saved most recently. Therefore the variable just output (C) is the second operand for the exponentiation, so this operator can be output. So far our postfix string has the form

A B C↑

Before we encountered the exponentiation we were trying to find the second operand for the addition, so we return to looking for that operand. We have just read the multiplication operator and done nothing with it, so we must now compare it with the addition we have been saving. Since multiplication has a higher precedence than addition, the compound operand B↑C just processed must be the left operand for the multiplication rather than the right operand for the addition. Therefore we save the multiplication until we have found its right operand. As we continue reading the input string, we read and output the next variable D and come to the division operation. Since this has the same priority as multiplication, the last variable read (D) is the right operand of the multiplication. So the multiplication can be put into the output string to get

A B C↑D*

Now we are back to looking for the rest of the second operand of the addition. Since the division just read has higher precedence than the addition, it is saved. We read and output the operand E and then come to the end of the expression. This is signaled by a special mark, or is indicated by a count of the number of items in the string. In Table A13.1 it is indicated by the symbol △. We can view the end of the expression as an operator of lowest precedence. Hence we can now output the division just saved. It has operands B*C↑D and E. This leaves only the addition to be processed. Since it also has higher precedence than the end of the expression, we have finally found its second operand, and we can output it to get

A B C↑D*E / +

During this process, it was necessary to save various operators until their second operands had been processed. When a new operator had to be saved, we temporarily ignored the earlier ones until we had processed the most recent. Thus we should save these operators on a stack. Each new one to be saved is placed in the stack; we concern ourselves with only the operator on top of the stack.

The process just described handles any valid combination of binary operators and operands. However, it does not allow for parentheses or for unary operators. Parentheses can be handled easily once we understand how they fit into the hierarchy. Their purpose is to override the natural hierarchy in which multiplication takes precedence over addition, and so forth. Any subexpression enclosed within parentheses must be evaluated immediately: it then takes the place of an operand for an operator outside the parentheses. Thus if we have the expression A*(B+C−D) / E, we must realize that the parentheses enclosing the addition and subtraction operators give them a higher precedence than the multiplication and division operators. If we start processing this expression by the algorithm given above, we will first output the operand A, then put the multiplication operator into the stack. The process is illustrated in Table A13.2. If we ignore the left parentheses, we will output the operand B and then come to the addition operator. Somehow we must recognize that this has a higher precedence than the multiplication operator currently in the stack. We could do this if we had put a mark on top of the stack to show that a parenthesis had been read. In fact, this mark could be the parenthesis itself if we had not ignored it but had put it in the stack. Thus our rule is that *a parenthesis on top of the stack has a lower precedence than any other operator read from the input.*

However, when we first read the left parenthesis, we put it in the stack. Therefore, our rule must also provide that *a left parenthesis in the input string has a higher precedence than anything in the stack* and should be put into the stack like an operator.

If we follow this process, we will continue scanning the expression A*(B+C−D) / E and output the variables A, B, and C, by which time

TABLE A13.2 HANDLING PARENTHESES

Input String (Infix)	Output String (Postfix)	Stack (Top→)	Comments
A*(B+C−D)/E△			
*(B+C−D)/E△	A		
(B+C−D)/E△	A	*	
B+C−D)/E△	A	*(	(higher than*
+C−D)/E△	A B	*(	
C−D)/E△	A B	*(+	+ higher than (
−D)/E△	A B C	*(+	
−D)/E△	A B C+	*(	+ = −
D)/E△	A B C+	*(−	− higher than (
)/E△	A B C+D	*(−	
)/E△	A B C+D−	*(	) lower than −
/E△	A B C+D−	*	Matching parentheses discarded
/E△	A B C+D−*		/ = *
E△	A B C+D−*	/	
△	A B C+D−*E	/	
△	A B C+D−*E/		△ lower than /

the stack will contain *, (, and +. When the subtraction is encountered the addition on top of the stack is output, since it has the same priority as subtraction. Then the subtraction is put into the stack, since it is higher than the left parenthesis in the stack, and the variable D is output. Now we reach the right parenthesis in the input that closes up the parenthesized expression. This means that the subtraction in the stack must be output, since the last operand read (D) is its second operand. Now we are in a state in which the stack has a left parenthesis on top and the last entry read from the input string is a right parenthesis. Since the left parenthesis was saved while we were processing the part of the string within the parentheses, we must have finished processing this part of the string, and the output postfix string must now contain a translated form of that subexpression. Consequently we can discard the parenthesis on top of the stack and continue scanning the input string. Next we read the division operator, whose precedence is equal to that of the multiplication on top of the stack, so we output the multiplication operator followed by the variable E, and the division operator is placed in the stack. Finally the end-of-expression character △ (which has the lowest precedence of all) is encountered, causing the division to be removed from the stack and output, and the process to terminate.

Program A13.1 converts an input string of tokens representing an infix expression *without* parentheses into a Polish postfix expression. Syntax errors are detected when the string does not consist of alternating operands and operators. Initially, the special marker △ is pushed into the stack. It is assumed to have lower precedence than any operator,

Program A13.1 ***Translating infix to postfix—no parentheses***

```
POSTFIX: program
   The input is read, a token at a time. The output is a Polish postfix
   expression. Data types of tokens are not declared, although typically
   they would be represented as integers.
   logical OPERAND_EXPECTED,NEXT_INPUT_PROCESSED
      initialize stack to empty
      push △ into stack
      The △ is pushed into the stack as a bottom marker. It will match
      the end-of-string indicator, and should be given the same
      precedence.
      OPERAND_EXPECTED←true
      NEXT_INPUT_PROCESSED←true
      do while stack not empty
         | if NEXT_INPUT_PROCESSED
         | then input NEXT_INPUT
         | endif
         | NEXT_INPUT_PROCESSED←false
         | if OPERAND_EXPECTED
         |    | then
         |    |    if NEXT_INPUT is an operand
         |    |       | then
         |    |       |    output NEXT_INPUT to Polish string
         |    |       |    OPERAND_EXPECTED←false
         |    |       |    NEXT_INPUT_PROCESSED←true
         |    |       | else
         |    |       |    output 'SYNTAX ERROR'
         |    |       |    NEXT_INPUT_PROCESSED←true
         |    |       | endif
         |    | else
         |    |    if NEXT_INPUT is an operator
         |    |       | then
         |    |       |    case precedence of NEXT_INPUT is
         |    |       |       (lower,equal,higher) than precedence of
         |    |       |       top of stack
         |    |       |         | lower: pop stack and output to Polish
         |    |       |         |    string
         |    |       |         | equal: pop stack and output to Polish
         |    |       |         |    string
         |    |       |         | higher: push NEXT_INPUT into stack
         |    |       |         |   NEXT_INPUT_PROCESSED←true
         |    |       |         |   OPERAND_EXPECTED←true
         |    |       |         | endcase
```

```
|  |        else
|  |            output 'SYNTAX ERROR'
|  |            NEXT__INPUT__PROCESSED←true
|  |        endif
|  endif
enddo
endprogram POSTFIX
```

and to match the precedence of the △ marker at the end of the string. It will be output to the Polish postfix expression when the translation has been completed. The logical variable OPERAND__EXPECTED indicates whether the next input token is expected to be an operand or operation.

Program A13.1 can be modified to handle parentheses by changing the first else clause to

```
else
    if NEXT__INPUT='('
        then push NEXT__INPUT into stack
        else output 'SYNTAX ERROR'
        endif
    NEXT__INPUT__PROCESSED←true
```

and the cases in the case statement to

```
lower:
    if top of stack='('
        then
            output 'SYNTAX ERROR, TOO MANY LEFT
              PARENTHESES'
            pop stack
        else
            pop stack and output to Polish string
        endif
equal:
    if top of stack='('
        then
            pop stack
            NEXT__INPUT__PROCESSED←true
        else
            pop stack and output to Polish string
        endif
higher:
    if NEXT__INPUT=')'
        then
```

```
            output 'SYNTAX ERROR, TOO MANY RIGHT
              PARENTHESES'
        else
            push NEXT__INPUT into stack
            OPERAND__EXPECTED←true
        endif
    NEXT__INPUT__PROCESSED←true
```

When an operand is expected and a left parenthesis is read, it is pushed into the stack and another operand sought. When a right parenthesis is read as an operator, it has the same precedence as a left parenthesis in the stack, a higher precedence than the bottom-of-stack marker △, and a lower precedence than anything else in the stack. This allows us to detect the cases when the next input and top of stack are one of the three pairs [△, (], [), △], and [), (]. The first two are syntax errors; the third is valid, and causes the input and top of stack to be discarded.

Problems

1. Convert the following infix expressions into trees and into postfix and prefix expressions:
 *a. A+B*C / D
 b. −A−C*D+B↑E
 c. (P+Q / R) / (S−T)
 d. X↑Y↑(−Z)

2. Convert the following prefix expressions into trees and into infix and postfix expressions:
 a. +−A B C D
 b. +m A*B m C
 c. ↑+P / Q R−S T
 d. ↑X+*Y Z W

3. Convert the following trees into prefix, postfix, and infix expressions:
 *a.

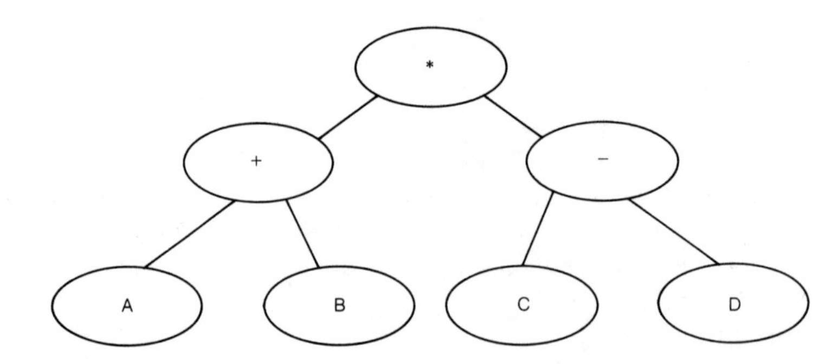

b.

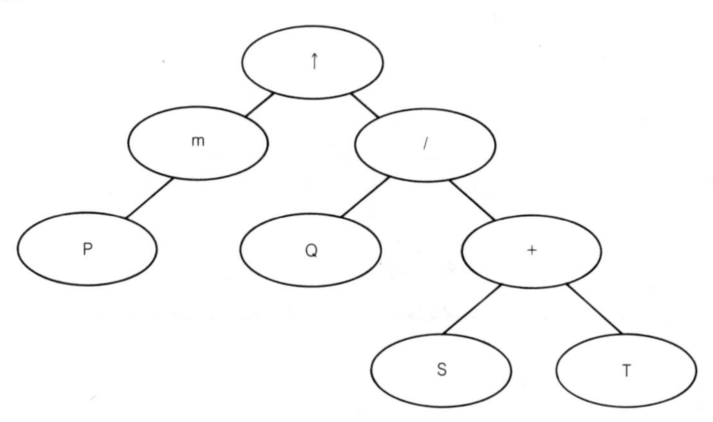

4. What is the value of the following postfix expressions?
   *a. 5.1 3 m+
   b. 4.8 m 1.3 2−
   c. 1.2 1.3+1.4 1.1+*
   d. 2 4.5 1.5 / 2*6+−
5. Write a program to input and convert a prefix expression into a tree. (Note that a binary tree can be used, so storage of the tree structure can be simple.) For the program, you can assume any suitably simple form of input coding. For example, positive integers can be used for operands and negative integers for operators.
*6. Write a program to convert the tree structure from Problem 5 into postfix form and to print it in that form.
7. Assume that a postfix expression is stored in the array K in the following way. Each item occupies one location. If an item is an operation, it is a negative integer with one of the following values:

+	−1
−	−2
*	−3
/	−4
↑	−5
m	−6

If an item is an operand, it is a positive integer J such that the value of the data is in A(J). Write a program to evaluate postfix expressions. Assume that the postfix string is terminated by the integer 0.
*8. Describe the changes necessary in Program A13.1 to handle the unary minus operator.

Chapter

A14

Solution of Nonlinear Equations

Many scientific and engineering applications require the solution of *nonlinear equations* that cannot be solved directly by arithmetic operations. Examples include second- and higher-order polynomials, such as

$$x^3 - x + 4.7 = 0$$
$$x^4 + 2x^2 - x + 4 = 0$$

and equations involving nonlinear functions, such as

$$\sin(x) - 0.5x + 1.3 = 0$$

Such an equation may have one real solution, none, or many. These three cases are illustrated in Figures A14.1 to A14.3, which show the graphs of the left-hand sides of the three equations above.

We will discuss several methods that can be applied to the problem of finding solutions to $f(x) = 0$, where $f(x)$ is some nonlinear function of x. We call such solutions *zeros* of $f(x)$. Most methods of solution require only that we can evaluate $f(x)$ for any value of the argument x. We can immediately see one source of error in the answer—rounding error in

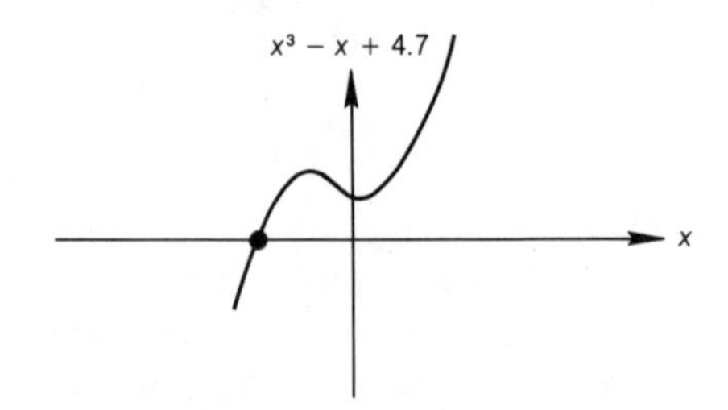

Figure A14.1 One real solution

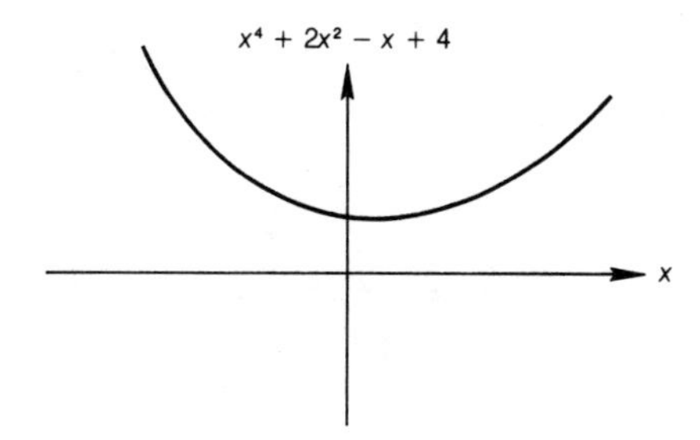

Figure A14.2 No real solution

evaluating $f(x)$. Since we cannot evaluate f exactly, the "graph" of $f(x)$ is effectively a "thick line." If we try to evaluate f with the unknown rounding errors, we may get any point inside this thick line. We want to know where the x-axis crosses the graph, but we can only find the *region* in which the x-axis lies within the "thick line," as shown in Figure A14.4. It is evident that the smaller the slope of the graph near the origin, the larger the region of intersection and hence the larger the possible error. Whatever method of solution we use, we cannot hope to get more accuracy than that represented by this region of intersection.

The method itself can introduce additional errors. For example, the method of bisection discussed in Chapter A2 can be used if $f(x)$ is continuous and we are given two values of x for which $f(x)$ has opposite signs. This guarantees that we start with an interval that contains at least one root. By making this interval successively smaller by factors of two, we can reduce it to any desired size within the precision of the arithmetic used. The length of this interval can cause an additional error, as shown in Figure A14.5.

All the methods we will discuss for the solution of nonlinear equations are iterative methods, like the method of bisection. In any iterative method, we have to decide when to stop the iteration. The method of bisection has the desirable property that it gives upper and lower bounds

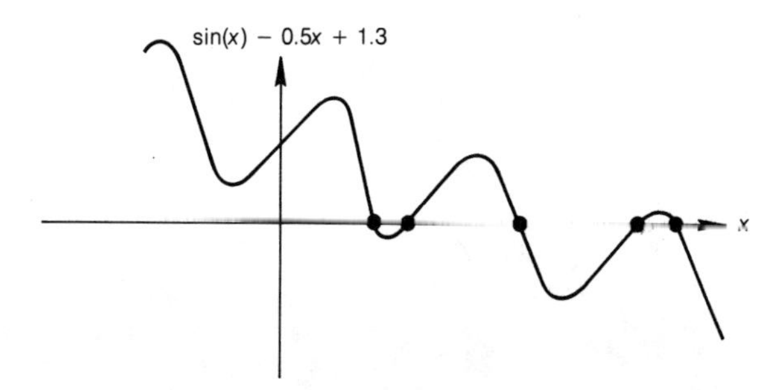

Figure A14.3 Many real solutions

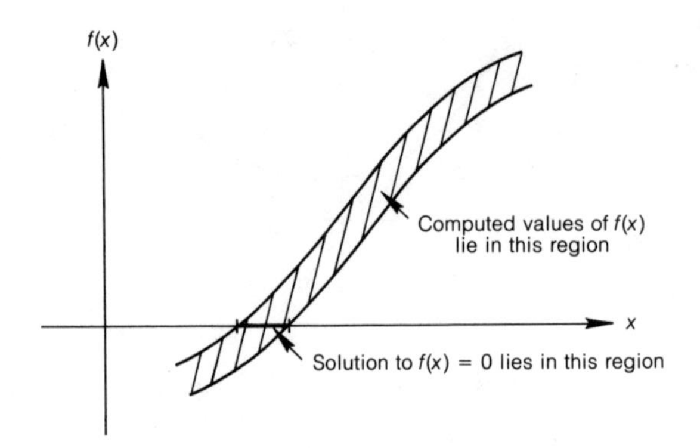

Figure A14.4 Effect of errors in evaluating $f(x)$

on the solution, so that we have a bound on the error introduced by the method (that is, the truncation error). We must realize that we cannot make this bound arbitrarily small because of the finite precision of the computer: we cannot reduce the interval between the upper and lower bounds below the minimum interval representable in the machine. Suppose we have a point L such that $f(L) < 0$ and a point U such that $f(U) > 0$, and suppose that U is the next larger number than L in the machine representation. The method of bisection calculates the midpoint of L and U as $(U + L)/2$. However, to machine precision, the value of this expression is either the point L or the point U. Therefore the next iteration of the method will give the same interval L to U as before, so there is no reason to continue. This is the closest we can get by any method of solution, although usually we will want to stop the iteration earlier. *In most problems of this sort, the rounding error in the machine* (*that is, the size of the smallest interval representable*) *is much less than the desired precision of the solution.*

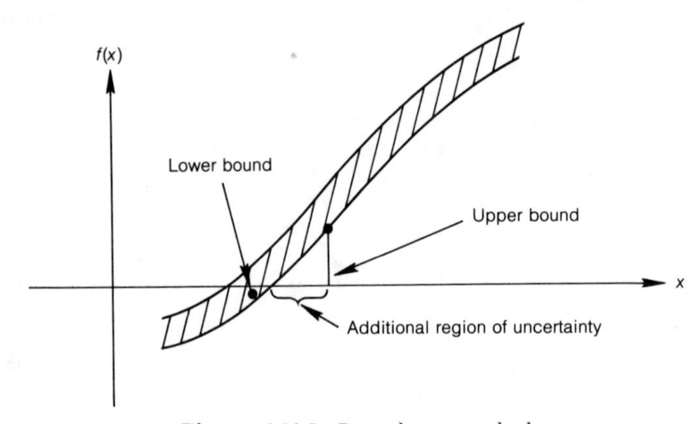

Figure A14.5 Error due to method

A14.1 REGULA FALSI METHOD

In the method of bisection, we evaluate the function but only use information about the sign. Suppose we try to use the size of the function as well. If the graph of the function is fairly straight, we can draw a straight line between the lower and upper points, and use this line to guess the point *P* where the function crosses the x-axis. (See Figure A14.6.) The point *P* is, we hope, a better approximation to the zero of the function than is the midpoint of the interval. As in the method of bisection, we form a new, smaller interval containing the solution. To do this we evaluate the function at the new point and decide on the basis of the sign which endpoint to replace with the new point *P*. In

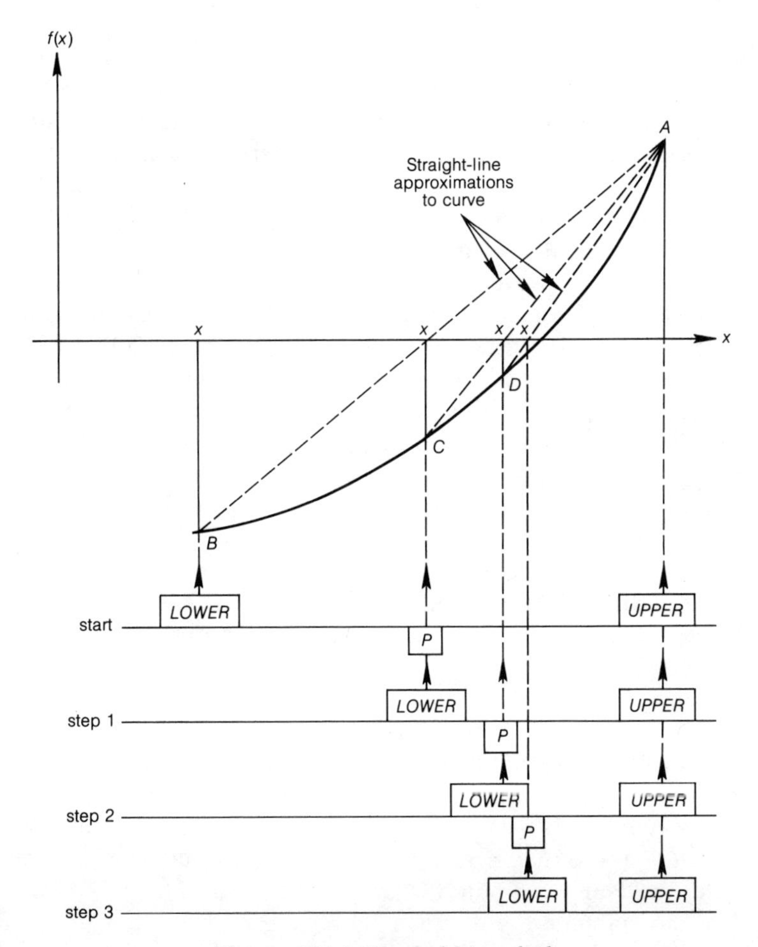

Figure A14.6 Regula falsi method

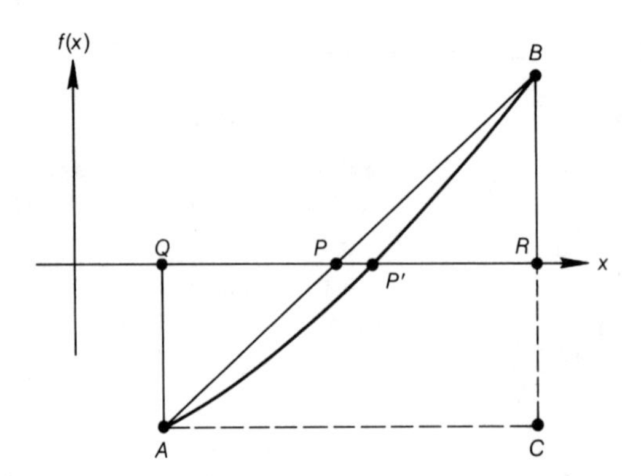

Figure A14.7 Using slope in stopping criterion

Figure A14.6, we see that the point P replaces the lower endpoint, because the sign of the function is negative at both points. This means that there is known to be a solution between P and the upper endpoint. This technique is called the *regula falsi method,* or *method of false position.*

The regula falsi method is guaranteed to give a solution, because we start with an interval known to contain a solution and always maintain such an interval. However, it may not give us upper and lower bounds that are close, as is demonstrated in Figure A14.6. In that example, the point P always replaces the lower bound, so the point *LOWER* slowly approaches the solution while the point *UPPER* never moves.

When should we stop this iteration? We could continue until a new interval is exactly the same as the previous one—but this might take an impossibly long time. We could continue until the value of $f(x)$ is less than some desired precision ε. This will enable us to solve the problem "find an x such that $|f(x)|$ is less than ε." It says nothing about how close the solution x is to a zero of $f(x)$. Since the relationship of the size of $f(x)$ to the distance of x from a zero of $f(x)$ depends on the slope of $f(x)$, we could use the slope as our stopping criterion, as shown in Figure A14.7.

If we approximate the function by the straight line AB between the two points A and B, then the slope of the function is approximated by BC/CA or by QA/QP. Suppose we have just computed the new point Q and want to know whether this is close enough to the solution P'—that is, whether it is within ε of P'. In other words, we want to know whether QP' is less than ε. Since we do not know where P' is, we assume it is close to P and compute QP.

We can compute the distances QA, BC, and CA. They are $-f(q)$, $f(r) - f(q)$, and $r - q$, respectively, where r and q are the x values of

R and Q, respectively. From the two forms for the slope we have

$$\frac{BC}{CA} = \frac{QA}{QP}$$

or

$$\frac{f(r) - f(q)}{r - q} = \frac{-f(q)}{QP}$$

That is,

$$QP = -f(q)\frac{r - q}{f(r) - f(q)}$$

We can thus calculate QP and find out whether it is less than ε. If it is, we can stop the iteration. Notice that the following step in the iteration would be to find the point P by approximating the function by the straight line AB. Thus QP is the amount by which we will change the point Q to get the next point P. Hence we are saying that the stopping criterion is to quit the iteration when the change in the estimated solution is less than our desired error ε.

There is no "best" or guaranteed criterion for stopping the iteration. The method just described is the most common, but it can give very poor answers, as the example in Figure A14.8 shows. Because the curve is not close to the straight line approximation AB, the point P, although close to Q, is very far from the solution W.

This example also shows that the regula falsi method can be slower than the bisection method. In Figure A14.8 the bisection method would reduce the interval from QR to MR in one step, whereas the regula falsi method reduces it only from QR to PR. However, both methods are guaranteed to find a solution eventually, since, within rounding

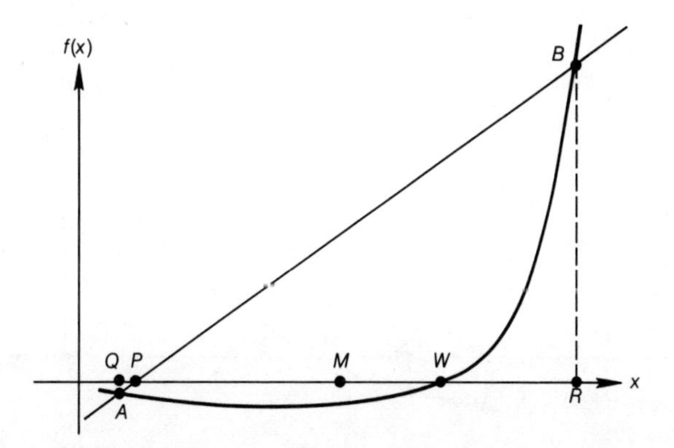

Figure A14.8 Poor convergence in regula falsi method

errors, the upper and lower bounds always enclose a solution. We say that the bisection and regula falsi methods *converge* to a solution: that is, the computed points get closer to the solution as the iteration proceeds.

A14.2 CHORD METHOD

As we have seen, the regula falsi method has the drawback that one of the endpoints may remain a long way from the solution. The function is usually approximated more accurately by a straight line between two points if the points are close together. For this reason, it is worth considering a method similar to the regula falsi method, but in which the last *two* points calculated are kept, as shown in Figure A14.9. The initial pair of points is x_0 and x_1. A straight-line approximation to the function is drawn between $f(x_0)$ and $f(x_1)$, intersecting the x-axis at x_2. The new pair of points is thus x_1 and x_2. The straight line between $f(x_1)$ and $f(x_2)$ intersects the x-axis at x_3, so that the next pair of points is x_2 and x_3. This process continues until some x_n and x_{n+1} are sufficiently close together.

This is called the *chord method*, or *secant method*, because the straight-line approximation is a *chord*, or *secant*, of the graph. Although it is usually faster than the bisection and regula falsi methods, the chord method no longer has the property of convergence, because we cannot guarantee that a solution lies between the currently retained pair of points. An example of nonconvergence is shown in Figure A14.10. The straight-line approximation between the points x_1 and x_2 is parallel to the x-axis, so it does not intersect the axis anywhere. When this happens, we have to "guess" another point and try again. Computer implementations of this method usually try first to use one of the previous points,

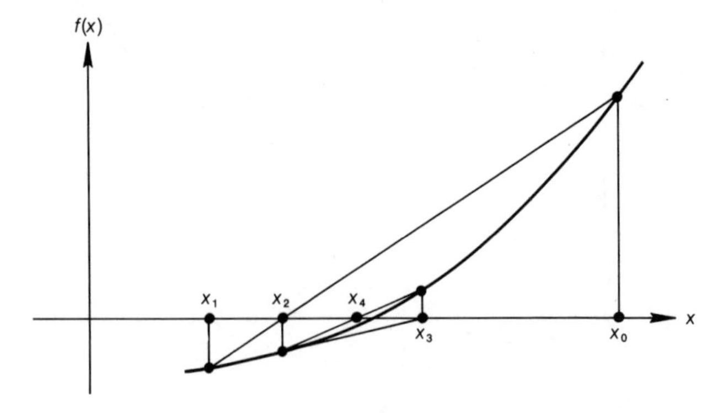

Figure A14.9 Chord method

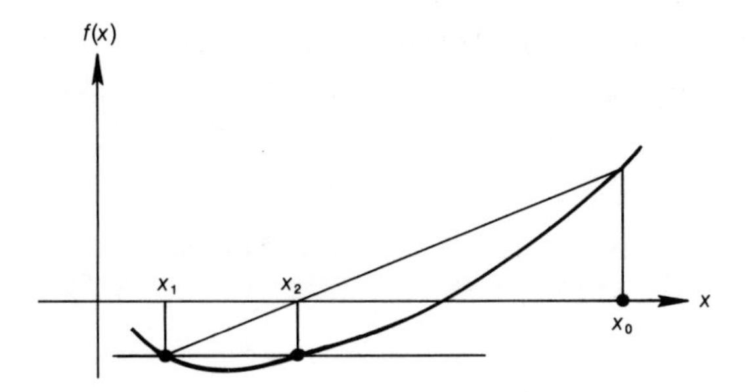

Figure A14.10 Nonconvergence in chord method

as in the regula falsi method, or, if that also fails, to move a given fixed amount in alternating directions. However, since convergence cannot be guaranteed, a computer program for the chord method must count the number of iterations attempted and quit if too many are used.

A14.3 NEWTON-RAPHSON METHOD

In the chord method we use two previous points to calculate a straight-line approximation to the function. If we know only one point P, at x_0 on the graph, we can calculate a second point P_1 at $x_0 + \Delta$, as shown in Figure A14.11, and pass a straight line through these two points to compute a new approximation x_1. The same process can then be repeated to get the point x_2. This might be called the *quasi-Newton method.*

The line through points P and P_1 in Figure A14.11 is used as an approximation to the curve. As Δ is made smaller, P_1 approaches P and the line through P and P_1 approaches the tangent to the curve at P, as

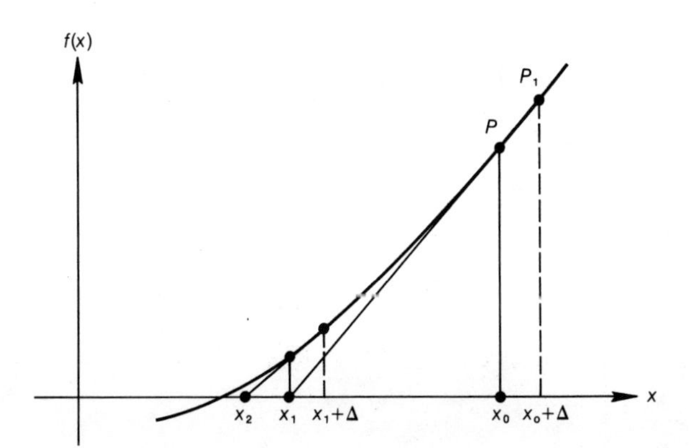

Figure A14.11 Quasi-Newton method

shown in Figure A14.12. In the *Newton-Raphson method*, or *Newton's method*, the tangent itself is used as the approximating line. If the slope of the tangent at point P is S, we have

$$S = \frac{PA}{PB} = \frac{f(x_0)}{x_0 - x_1}$$

Therefore

$$x_0 - x_1 = f(x_0)/S$$

or

$$x_1 = x_0 - f(x_0)/S$$

The general step is given by

$$x_{n+1} = x_n - f(x_n)/S_n$$

where S_n is the slope of the tangent at x_n. To use this method we must be able to calculate both the function $f(x)$ and the slope of its tangent. (Readers who have had calculus will recognize this as the *derivative* of the function.) Consequently the Newton-Raphson method is mainly used for simple functions, such as polynomials, for which methods of calculating the tangent are simple.

A particularly important case is the equation $x^2 - a = 0$. A solution of this equation is $\sqrt{a}$. The slope of the tangent* is 2x. In this case, the Newton-Raphson method gives

---

*Students who have not had calculus can find the slope as follows: The tangent is the limiting case of a line passing through two points on a curve as the two points get closer and closer. Consider the figure below and the line through the two points P at x and Q at $x + \Delta$.

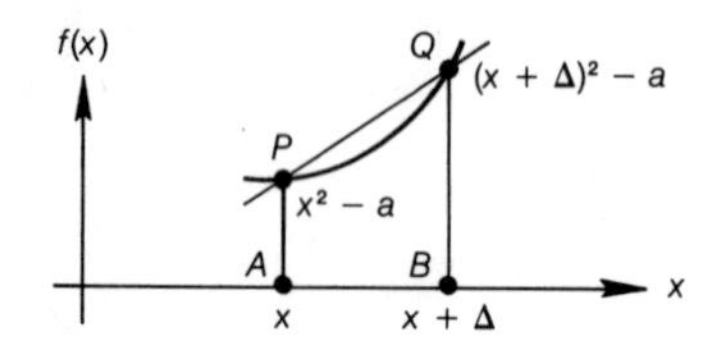

The slope of this line is

$$S = \frac{QB - PA}{B - A} = \frac{(x + \Delta)^2 - x^2}{(x + \Delta) - x}$$

$$= \frac{x^2 + 2\Delta x + \Delta^2 - x^2}{x + \Delta - x}$$

$$= \frac{2\Delta x + \Delta^2}{\Delta}$$

$$= 2x + \Delta$$

As Δ gets smaller, the point B approaches the point A and the slope approaches 2x.

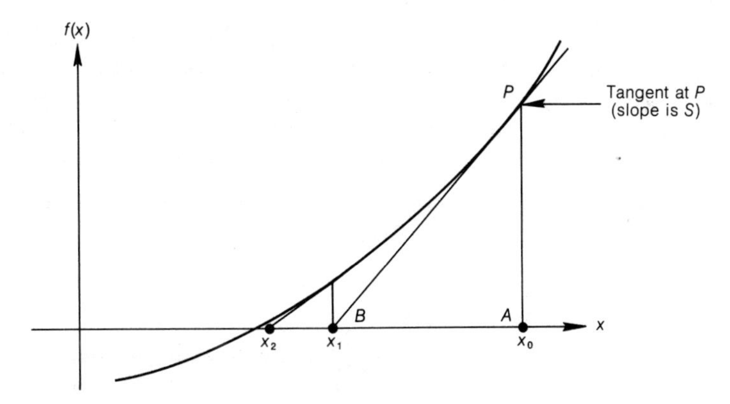

Figure A14.12 Newton-Raphson method

$$x_{n+1} = x_n - \frac{f(x_n)}{S_n}$$

$$= x_n - \frac{x_n^2 - a}{2x_n}$$

$$= \frac{(x_n + a/x_n)}{2}$$

We can show that this particular iteration will always converge, because we have

$$x_{n+1} - \sqrt{a} = \frac{x_n^2 + a}{2x_n} - \sqrt{a}$$

$$= \frac{(x_n - \sqrt{a})^2}{2x_n}$$

Therefore $x_{n+1} - \sqrt{a}$ has the same sign as x_n. Assume that we start with a positive x_0 and that we take $\sqrt{a}$ to be positive. Then $x_n \geq \sqrt{a} \geq 0$ for all $n \geq 1$. Furthermore, the error in x_{n+1} is

$$x_{n+1} - \sqrt{a} = \frac{(x_n - \sqrt{a})^2}{2x_n} = (x_n - \sqrt{a})\frac{x_n - \sqrt{a}}{2x_n}$$

For $n \geq 1$, $x_n \geq \sqrt{a}$, so that $x_n - \sqrt{a} \leq x_n$. Hence

$$\frac{x_n - \sqrt{a}}{2x_n} \leq \frac{1}{2}$$

Substituting this in the previous equation, we get

$$x_{n+1} - \sqrt{a} \leq \frac{1}{2}(x_n - \sqrt{a})$$

That is, the error in x_{n+1} is no more than 50% of the error in x_n. In fact, the Newton-Raphson method converges very much more rapidly than this. It is one of the fastest methods available for those problems for which it converges, but it is not guaranteed to converge for all problems. (What happens if the slope of the tangent is zero near the solution?)

Problems

***1.** Write a subroutine for the chord method. The parameters to the subroutine should be A, B, X, F, and E, where A and B are the two initial points, F is the function whose zero is to be found, and E is an error tolerance. The iteration should stop when the change in the computed solution in one iteration is less than E. X should be left containing the solution.

2. Consider your solution to Problem 1. Since the stopping criterion may never be satisfied, the subroutine may never terminate. Add another logical parameter whose value is true if a solution is found, but whose value is false if the stopping criterion is not satisfied after 50 iterations.

***3.** Sketch a function for which the Newton-Raphson method will give a sequence of approximations x_n, each of which is further from the zero than the last.

Chapter

A15

Best Approximation: Least Squares and Chebyshev

There are many situations in which the scientist or engineer needs to approximate a function. A scientist may have taken some measurements and wish to derive a simple function that approximates the observed results. Alternatively, an engineer may know a function mathematically or experimentally and wish to compute its values approximately. Series and interpolatory methods for this have been discussed in Chapters A5 and A8. In this chapter, we will develop two important techniques for approximation that can be used for either of these tasks.

A15.1 LEAST-SQUARES APPROXIMATION

Least-squares approximation can be used to derive a simple function that approximates the observed results of an experiment. This application frequently arises when a scientist hypothesizes a formula governing a process, and then wishes to measure the coefficients in that formula. Suppose, for example, that we wish to measure the acceleration due to gravity, g, by dropping an object and measuring its velocity after a given period of time. Since the velocity of an object starting from rest at time 0 is gt after t seconds, g can be found by measuring the velocity after one second. If this is found to be 980 centimeters per second, then we can conclude that the acceleration due to gravity, g, is 980 centimeters per second per second—or can we? Clearly there will be experimental error. Consequently, we may repeat our experiment a number of times and average the results, hoping, thereby, to reduce the experimental error. Before deciding the best way average the results to reduce the amount of error, we need to consider the nature of the errors.

The first and most obvious cause of experimental error is *measurement error*. Because instruments can only be read with limited accuracy, typically between 0.1% and 1.0%, we can only expect two or three decimal digits of precision in the measurements. These errors are similar to rounding errors in computation: they are random and unpredictable, although we do have some idea of their worst-case size.

The second cause of error is consistent deviations in experimental conditions. For example, in the object-dropping experiment, the device that releases the object at time 0 may give it a small initial velocity u. In that case, the velocity after t seconds will be $u + gt$ instead of gt. This constant *bias*, u, will make g appear larger if u is positive or smaller if u is negative. Consequently, rather than just calculate g, we should also attempt to calculate u. That is, we hypothesize a formula $v = u + gt$ connecting our experimental observations of time t and velocity v, and calculate both g and u.

How can we do this? One way is to take two measurements, say v_1 after time t_1 and v_2 after time t_2, and solve for u and g in the equations

$$v_1 = u + gt_1$$
$$v_2 = u + gt_2$$

These are two equations in two unknowns and so, if t_1 and t_2 are different, can be solved uniquely. Unfortunately, we still have the problem of measurement errors, so we would like to take many measurements at different times and use the information to make a "best" estimate of the values of u and g.

Figure A15.1 shows a graph of $v = u + gt$ for particular values of u and g, along with a set of measurements that might be taken in an experiment. Some of the measurements are very close to the assumed function, so we can find no reason to disbelieve the graph on the basis of these measurements. However, some of the measurements, such as v_{n-1} and v_n, are a long way from the line representing the function, leading us to suspect that u and g should be changed so that the function is closer to those measurements. What we wish to do is choose u and g so that the function is not too far away from any of the measured points, since we are prepared to accept small deviations but disbelieve large ones. Consequently, in trying to estimate how far the measurements are from the function, we want to give more weight to points further away. If we take the sum of the squares of the errors, that is,

$$D = (v_1 - u - gt_1)^2 + (v_2 - u - gt_2)^2 + \ldots + (v_n - u - gt_n)^2$$

we get a number D that is never negative, and becomes larger if any measured point is moved further away from the graph of the function. Furthermore, the contribution to D is much larger for points that are further away than for points that are close. For example, if a point is

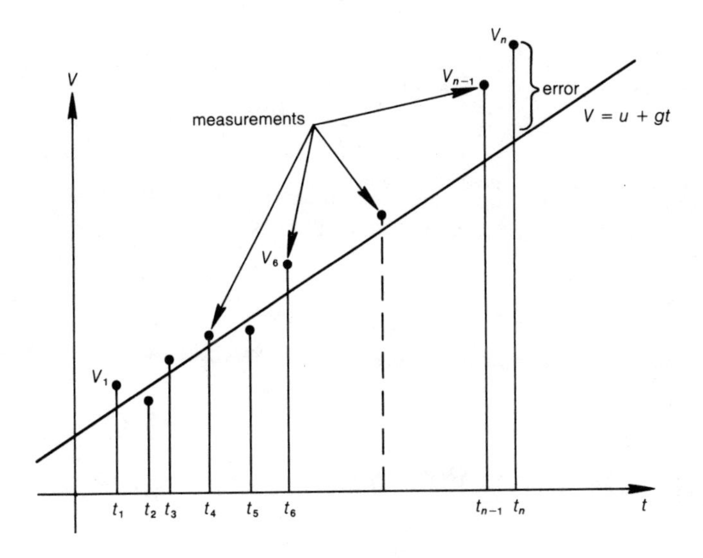

Figure A15.1 Measurements and hypothesized function

twice as far from the graph of the function as another point, it contributes four times as much to D, because of the effect of the square. Hence it seems very reasonable to pick u and g so as to make D as small as possible. This process is called finding a *least-squares fit,* in this case, a *linear least-squares fit,* because the approximating function, $u + gt$, is linear in the variable t. (The use of a least-squares fit as the most reasonable approximation can be justified on other, more technical grounds as well.)

The least-squares fit has one very useful feature: it is easy to calculate the coefficients that make D as small as possible. The derivation we will give uses simple calculus; the same result can be derived without calculus, but it takes considerably longer. Readers who do not know calculus can skip the derivation.

The sum of the square errors is

$$D = \sum_{i=1}^{n} (v_i - u - gt_i)^2$$

The values of u and g that minimize this expression are found by differentiating D with respect to u and g separately, and equating the derivatives to zero. Thus we get

$$\frac{\delta D}{\delta u} = 0 = -2 \sum_{i=1}^{n} (v_i - u - gt_i)$$

and

$$\frac{\delta D}{\delta g} = 0 = -2 \sum_{i=1}^{n} (v_i - u - gt_i)t_i$$

These equations can be rewritten as

$$u \sum_{i=1}^{n} 1 + g \sum_{i=1}^{n} t_i = \sum_{i=1}^{n} v_i$$

and

$$u \sum_{i=1}^{n} t_i + g \sum_{i=1}^{n} t_i^2 = \sum_{i=1}^{n} v_i t_i$$

If we write

$$T_0 = \sum_{i=1}^{n} 1 = n; \; T_1 = \sum_{i=1}^{n} t_i; \; T_2 = \sum_{i=1}^{n} t_i^2$$

$$V_0 = \sum_{i=1}^{n} v_i; \; V_1 = \sum_{i=1}^{n} v_i t_i$$

the equations determining u and g are

$$uT_0 + gT_1 = V_0$$
$$uT_1 + gT_2 = V_1$$

which can be solved for u and g. As long as not all the values of t_i are identical, it can be shown that there is a unique solution for u and g.

Least-squares fit is not restricted to linear approximations. For example, we could have chosen to measure g by finding out how far an object fell in t seconds, with $t \neq 1$. The formula for the distance fallen d is

$$d = s + ut + gt^2/2$$

where s is the initial distance, u the initial velocity, and g the acceleration due to gravity. If a series of measurements t_i and d_i are taken, we can choose s, u, and g so as to minimize

$$D = \sum_{i=1}^{n} (d_i - s - ut_i - gt_i^2/2)^2$$

If we do this, we obtain

$$sT_0 + uT_1 + \frac{g}{2}T_2 = D_0$$

$$sT_1 + uT_2 + \frac{g}{2}T_3 = D_1$$

$$sT_2 + uT_3 + \frac{g}{2}T_4 = D_2$$

where

$$T_0 = \sum_{i=1}^{n} 1 = n; \; T_1 = \sum_{i=1}^{n} t_i; \; T_2 = \sum_{i=1}^{n} t_i^2; \; \ldots ;$$

$$D_0 = \sum_{i=1}^{n} d_i; \; D_1 = \sum_{i=1}^{n} d_i t_i; \; D_2 = \sum_{i=1}^{n} d_i t_i^2$$

These three equations can be solved for s, u and g.

Example A15.1 ***Linear Least-Squares Fit***

Suppose we expect a measurement of the temperature y of a body to vary with time t according to the equation $y = a + be^{-t}$. We have the following measurements at various times t_i:

i	t_i	y_i
1	0	1.26
2	1	1.10
3	2	0.895
4	4	0.758
5	8	0.761

We wish to find the coefficients a and b.

It seems that this is not a straight-line approximation, so our least-squares method is of no value. However, we can change to a new variable $x = e^{-t}$, so that $y = a + bx$. This formula is linear and we can do a least-squares fit, so we proceed to form

$$X_p = \sum_{i=1}^{5} x_i^p$$

$$Y_p = \sum_{i=1}^{5} y_i x_i^p.$$

The results are shown in Table A15.1.

We substitute these values in

$$X_0 a + X_1 b = Y_0$$
$$X_1 a + X_2 b = Y_1$$

to get

$$5a + 1.52187b = 4.774$$
$$1.52187a + 1.15399b = 1.79994$$

TABLE A15.1 VALUES OF x_i, y_i, x_i^2, AND $y_i x_i$

i	t_i	$x_i = e^{-t_i}$	y_i	x_i^2	$y_i x_i$
1	0	1.00000	1.26	1.00000	1.26000
2	1	0.36788	1.10	0.13534	0.40467
3	2	0.13534	0.895	0.01832	0.12113
4	4	0.01832	0.758	0.00033	0.01389
5	8	0.00033	0.761	0.00000	0.00025
Sum		1.52187	4.774	1.15399	1.79994
		X_1	Y_0	X_2	Y_1

These equations have the solution

$$a = 0.80197$$
$$b = 0.50213$$

so the least-squares fit for this data is

$$y = 0.80197 + 0.50213e^{-t}$$

Least-squares fit can also be used to approximate functions. In Chapter A8 we used linear interpolation to approximate a function measured experimentally. Alternatively, we can approximate a function by a polynomial and determine the coefficients of the polynomial by the least-squares method. Then numerical approximations to the function can be obtained by evaluating the polynomial.

A15.2 CHEBYSHEV APPROXIMATION

Suppose we are asked to write a subprogram that will approximate sin(x) to limited precision for small values of x and will operate very rapidly (presumably because it is to be used frequently in a calculation). We can start with the power series for the sine and discard all but a few terms. Suppose we discard all but the first term. We then have the approximation

$$\sin(x) = x$$

If the maximum error that can be tolerated is e, the largest value of x for which this approximation is sufficiently accurate is the value x_0 for which $x_0 = \sin(x_0) + e$, as shown in Figure A15.2. However, if we choose another linear approximation, say

$$\sin(x) = ax$$

where a is less than one, we may be able to extend the range of x for which the approximation is within e to all x less than some larger x_1,

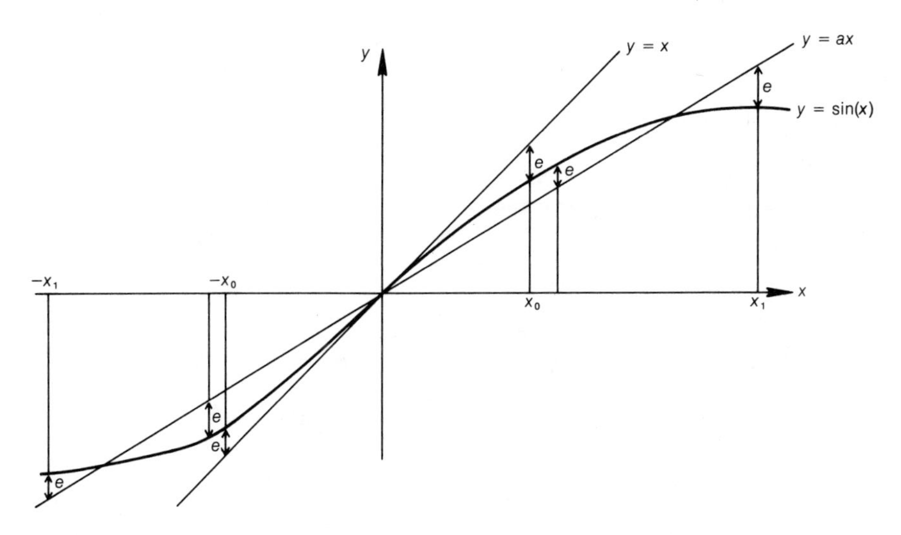

Figure A15.2 Linear approximations to sin(x)

again as shown in Figure A15.2. The desired property of the straight line $y = ax$ in relation to the function sin(x) over the interval $(-x_1, x_1)$ is that, of all linear functions, its maximum deviation from sin(x) is the smallest. That is, if we take any other linear function and calculate the largest value of [linear function − sin(x)] over the interval $(-x_1, x_1)$, we will get a number larger than e. For this reason, we call this the *minimax* or *Chebyshev* approximation to sin(x) over this interval.

Chebyshev approximations, named for the Russian mathematician Pafnuti Lvovich Chebyshev, are used in many computer routines. They are not limited to linear polynomials, but can be used to approximate nonlinear functions, such as the sine function in our example: the polynomial of a certain fixed degree is found whose maximum deviation from the sine over the interval in question is minimum.

Linear Chebyshev approximations are not too difficult to calculate for many problems. For example, if we wish to approximate sin(x) over the interval $(0, \pi/4)$ with a linear function, we will choose the function shown in Figure A15.3. The error at $x = 0$ is some number d, the error at the extreme on the other side (P in the figure) is $-d$, and the error at $\pi/4$ is d again. That is, the worst-case error occurs at three points, and the sign of the error at those points oscillates. It can be shown that the Chebyshev approximation always has this property for linear approximations. Generally, the Chebyshev approximation by an nth degree polynomial has $n + 2$ extreme errors with alternating signs.

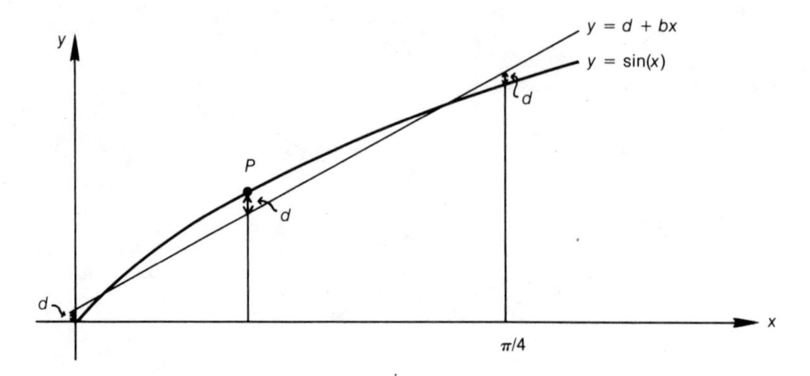

Figure A15.3 Chebyshev approximation to sin(x)

Problem

1. Suppose the following measurements are taken in an experiment:

x	1	2	3	4
y	4	1	0	−1

Find M and B in $y = Mx + B$ by a least-squares technique.

Appendix

Answers to Selected Problems

Chapter A3

3. a. Reverse ordering (5, 4, 3, 2, 1) leads to the worst case. The inner loop is executed 1 + 2 + 3 + 4 = 10 times.

b. Selection sort: 10 (outer loop executed 5 times, two assignments on each pass)
Bubble sort: 30 (inner loop executed 6 * 5 / 2 = 15 times, two assignments on each pass)
Insertion sort: 15 (inner loop executed 6 * 5 / 2 = 15 times, one assignment on each pass)

c. The general cases are 2*(N−1), N*(N−1), and N*(N−1) / 2, respectively.

Chapter A4

2. a.

```
ADD: subprogram (DATA,PTR,N,AVAIL,NAME,END)
    The name NAME is added to the end of the list kept in DATA and
    PTR, starting in PTR(1) and ending at END. The free list starts in
    AVAIL.
    integer N,PTR(N),I,AVAIL,END
    character (20)DATA(N),NAME
        if AVAIL=0
            then output 'NO SPACE LEFT IN FREE LIST'
            else
                I←AVAIL
                AVAIL←PTR(I)
                PTR(I)←0
```

```
            DATA(I)←NAME
            PTR(END)←I
            END←I
        endif
    return
endsubprogram ADD
```

b. Modify subprogram DELETE to include the parameter END, and insert the following line before endif:

```
if PTR(I)=0 then END←I endif
```

5.

```
SORT__LIST: subprogram (DATA,PTR,N)
    The list in DATA and PTR, starting in PTR(1), is sorted by a form of
    insertion sort.
    integer N,PTR(N),I,TEMP__PTR,MOVE
    character (20)DATA(N)
        TEMP__PTR←PTR(1)
        PTR(1)←0
        do while TEMP__PTR≠0
            The list starting in PTR(1) is in ascending order. Add next
            element from list starting in TEMP__PTR to it.
            MOVE←TEMP__PTR
            TEMP__PTR←PTR(TEMP__PTR)
            I←1
            do while PTR(I)≠0 and DATA(PTR(I))<DATA(MOVE)
                I←PTR(I)
                enddo
            Put MOVE after I in sorted list.
            PTR(MOVE)←PTR(I)
            PTR(I)←MOVE
            enddo
        return
    endsubprogram SORT__LIST
```

Chapter A8

2.

```
ZERO: subprogram (X,Y,N)
    Value of ZERO is such that Y(ZERO) = 0 under the assumption that
    Y(X) can be approximated linearly from the tabulated values in X(N),
    Y(N), and that the signs of Y(1) and Y(N) are different. If Y(1) and
```

Y(N) have the same sign, an error message is printed and 0.0 is returned.

```
integer I,N
real X(N),Y(N),ZERO,XV=0.
    if Y(1)<0
       | then
       |     if Y(N)<0
       |        | then
       |        |     I←0
       |        | else
       |        |     do for I←2 to N while Y(I)<0
       |        |         enddo
       |        | endif
       | else
       |     if Y(N)≥0
       |        | then
       |        |     I←0
       |        | else
       |        |     do for I←2 to N while Y(I)≥0
       |        |         enddo
       |        | endif
       | endif
    if I=0
       | then
       |     output 'NO SIGN CHANGE BETWEEN Y(1) AND Y(N)'
       |     XV←0.0
       | else XV←X(I−1)+Y(I−1)*(X(I−1)−X(I))/(Y(I)−Y(I−1))
       | endif
    return (XV)
 endsubprogram ZERO
```

Chapter A9

2. The outline of Program A9.2 below shows each do statement followed by the number of times the body of that loop is executed.

```
do for I←1 to N
    N
    do for J←I+1 to N
        N − 1 + N − 2 + . . . + 1 = (N − 1)N / 2
    do for J←I to N+1
        N + 1 + N + . . . + 4 + 3 = (N − 1)(N + 4) / 2
        (Note that the if condition cannot be true when I=N.)
    do for J←I+1 to N+1
        N + N − 1 + . . . + 1 = N(N + 1) / 2
```

```
        do for J←I+1 to N
           N − 1 + N − 2 + . . . + 1 = (N − 1)N / 2
           do for K←I+1 to N+1
              (N − 1)N + (N − 2)(N − 1) + . . . + 1*2 =
              (N − 1)N(N + 1) / 3
     outer: do for J←N−1 to 1 by −1
        N − 1
        inner: do for K←J+1 to N
           1 + 2 + . . . + N − 1 = (N − 1)N / 2
```

4. b.

Code segment to unscramble values in A(I,N+1).

```
do for I←1 to N
   if ROW(I)≠I
      then
         TEMP←A(I,N+1)

         J←I
         do while ROW(J)≠I
            A(J,N+1)←A(ROW(J),N+1)
            K←J
            J←ROW(J)
            ROW(K)←K
            enddo
         A(J,N+1)←TEMP
         ROW(J)←J
      endif
   enddo
```

Chapter A10

1. Rounding error is error caused by the fact that, since the computer does not have infinite precision, the results of arithmetic operations may have to be approximated.

3. A given rounding error is proportional to the size of the number rounded. If this number is large, and another large number of about the same size is later subtracted from it, the size of the earlier rounding error will be larger relative to the difference.

7. Yes. The hypotenuse is $(A^2 + B^2)^{1/2}$. A small change in A or B will not change the answer by an amount greater than that change.

10. b. If COLUMN(K) contains I, then x_i is in A(K,N+1).

Chapter A11

3. Modify the TELLER subprogram by replacing the long if-then-else group with

```
Get first element from line K.
J←REMOVE(K)
if J≠0
    then
        TELTIM(I)←TIME+SERVTM(J)
        WAIT←WAIT+TIME−CUSTAR(J)
        WAITSQ←WAITSQ+(TIME−CUSTAR(J))↑2
        NCUSTS←NCUSTS+1
        BUSY(I)←true
    else
        BUSY(I)←false
    endif
```

Modify the ARRIVAL subprogram by replacing the outermost if-then-else group with

```
Find shorter line and add element to it.
J←1
if QLENGTH(2)<QLENGTH(1) then J←2 endif
I←ADDEND(J)
if I=0
    then
        output 'NO FREE SPACE LEFT, SIMULATION ABANDONED'
        TIME←1.0E50
    else
        New entry is at index I.
        ARRTIME←TIME+2.0+4.0*RANDOM()
        SERVTM(I)←1.0+9.0*RANDOM()
        CUSTAR(I)←TIME
        QLENGTH(J)←QLENGTH(J)+1
        if not BUSY(1)
            then call TELLER(1)
            else if not BUSY(2) then call TELLER(2) endif
            endif
    endif
```

Add the two new subprograms REMOVE and ADDEND:

```
REMOVE: subprogram (K)
    Remove first element of list starting in PTR(K) and ending in
    LEND(K). Value of REMOVE is index to removed element if there is
    one, else 0.
    (all global declarations from main program)
    integer REMOVE,K,I
        I←PTR(K)
        if I≠0
            then
                PTR(K)←PTR(I)
                PTR(I)←FREE
                FREE←I
                if PTR(K)=0 then LEND(K)←K endif
            endif
        return (I)
    endsubprogram REMOVE
```

```
ADDEND: subprogram (J)
    Add new element to end of list ending in LEND(J). Index of new
    element is value of ADDEND. If no free space, value of ADDEND
    is zero.
    (all global declarations from main program)
    integer ADDEND,J,I
        I←FREE
        if I≠0
            then
                FREE←PTR(I)
                PTR(I)←0
                PTR(LEND(J))←I
                LEND(J)←I
            endif
        return (I)
    endsubprogram ADDEND
```

Chapter A12

1. If the longest path is n, there will be n levels below the root. One node can be at the root, two nodes at the next level, four at the next, eight at the next, and so on. Thus the total number of nodes cannot exceed

$$1 + 2 + 4 + 8 + \dots + 2^{n-1} = 2^n - 1.$$

When the number of levels is 3, we find the following distribution of search times:

1	node takes 1	comparison
2	nodes take 2	comparisons
4	nodes take 3	comparisons

The average over the seven nodes is
$(1 \times 1 + 2 \times 2 + 4 \times 3)/7 = 2\ 3/7$
For four levels we get
$(1 \times 1 + 2 \times 2 + 4 \times 3 + 8 \times 4)/15 = 3\ 4/15$
For five levels we get
$(1 \times 1 + 2 \times 2 + 4 \times 3 + 8 \times 4 + 16 \times 5)/31 = 4\ 5/31$
For n levels we will get $n - 1 + n/(2^n - 1)$.

4. If all the words containing definitions are input before the words to be looked up, it is easiest to use a fast sort method for arrays, such as those discussed in Chapter A3. Then a fast binary search can be used on the array. In the second case, it is better to sort the data as it is input, so that a search can be done at any time. Therefore a binary tree may be preferable.

Chapter A13

1. a.

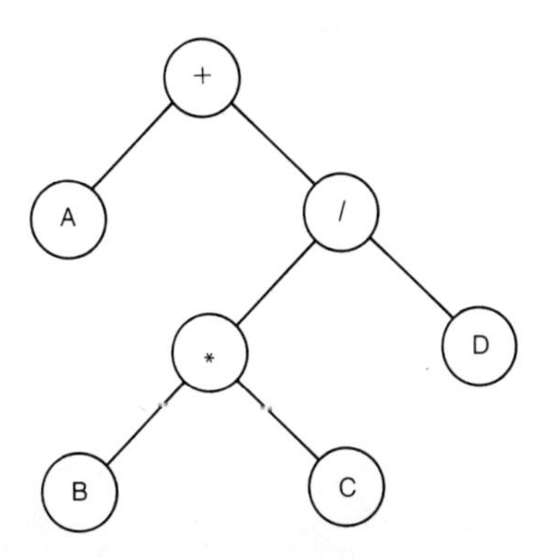

Postfix: A B C * D / +
Prefix: + A / * B C D

2. a.

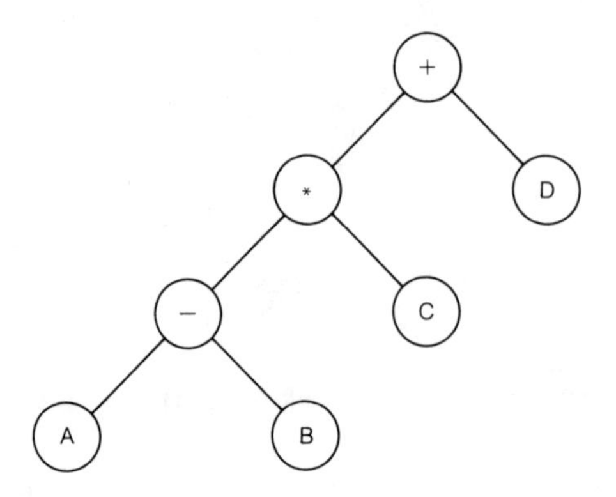

Infix: (A−B)*C+D
Postfix: A B−C*D+

3. a. Prefix: *+A B−C D
Postfix: A B+C D−*
Infix: (A+B)*(C−D)

4. a. +2.1
b. −7.4

6.

```
TREE__TO__POSTFIX: subprogram (ROOT,LEFT,DATA,RIGHT,N)
  The tree rooted in ROOT is converted to postfix form. A stack is
  used to save the path down into the tree (compare with Program
  A12.3). However, it is necessary to save a pointer to a node in the
  stack until both the left and right subtrees have been examined, so
  a positive value is stored while the left tree is being processed, and it
  is switched to a negative value while the right subtree is processed.
      integer ROOT,N,LEFT(N),DATA(N),RIGHT(N),PTR
      set stack to contain 0 as bottom marker
      PTR←ROOT
      do while PTR≠0 or stack not empty
          GO__LEFT: do while PTR≠0
              push PTR onto stack
              PTR←LEFT(PTR)
              enddo GO__LEFT
          pop stack to PTR
          do while PTR<0
              If node has already been visited twice, output DATA
              and pop up to a higher level.
              output DATA(−PTR)
              pop stack to PTR
              enddo
```

```
            if PTR≠0
                then
                    push −PTR onto stack
                    PTR←RIGHT(PTR)
                endif
            enddo
        return
    endsubprogram TREE__TO__POSTFIX
```

Recursion allows a much simpler statement of the program, as follows:

```
TREE__TO__POSTFIX: subprogram (ROOT,LEFT,DATA,RIGHT,N)
    The tree rooted at ROOT is output in postfix order recursively by
    printing each node in postfix order.
    integer ROOT,N,LEFT(N),DATA(N),RIGHT(N)
        if ROOT≠0
            then
                call TREE__TO __POSTFIX(LEFT(ROOT),LEFT,DATA,
                  RIGHT,N)
                call TREE__TO __POSTFIX(RIGHT(ROOT),LEFT,DATA,
                  RIGHT,N)
                output DATA(N)
            endif
        return
    endsubprogram TREE__TO __POSTFIX
```

8. Change the first else clause to

```
else
    if NEXT__INPUT='−'
        then push unary minus into stack
        else . . . previous contents of else clause . . .
        endif
```

Chapter A14

1.

```
CHORD: subprogram (A,B,X,F,E)
    Compute the solution of F(X)=0 using the chord method with
    starting values A and B. Stop when change in solution is less than E.
```

```
real A,B,X,F,E,X0,X1,F0,F1,CHANGE
    X0←A
    X1←B
    F0←F(X0)
    F1←F(X1)
    CHANGE←(X1−X0)*F1 / (F1−F0)
    do while CHANGE≥E
        X0←X1
        F0←F1
        X1←X1−CHANGE
        F1←F(X1)
        CHANGE←(X1−X0)*F1 / (F1−F0)
        enddo
    X←X1
    return
endsubprogram CHORD
```

3. The function $f(x) = x^{1/3}$, which has a zero at $x = 0$, has the property that the Newton-Raphson method diverges. In fact, it can be shown that $x_{n+1} = -2x_n$, as illustrated in the figure below:

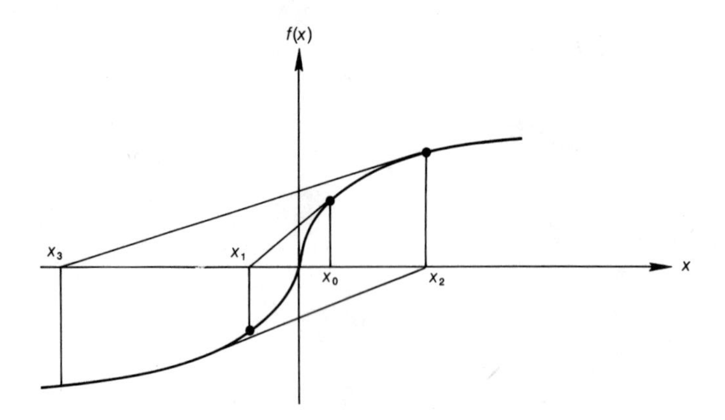

Index